Mathematical
Olympiad
in China (2023)

Problems and Solutions

Mathematical Olympiad Series

ISSN: 1793-8570

Series Editors: Lee Peng Yee *(Nanyang Technological University, Singapore)*
Xiong Bin *(East China Normal University, China)*

Published

Vol. 29 *Mathematical Olympiad in China (2023): Problems and Solutions*
edited by Bin Xiong (East China Normal University, China)

Vol. 28 *IMO Problems, Theorems, and Methods: Number Theory*
by Bin Xiong (East China Normal University, China) &
Gengyu Zhang (East China Normal University, China)

Vol. 27 *IMO Problems, Theorems, and Methods: Geometry*
by Tianqi Lin (Fudan University Affiliated High School, China) &
Bin Xiong (East China Normal University, China)

Vol. 26 *IMO Problems, Theorems, and Methods: Combinatorics*
by Guangyu Xu (East China Normal University, China) &
Zhenhua Qu (East China Normal University, China)

Vol. 25 *IMO Problems, Theorems, and Methods: Algebra*
by Jinhua Chen (East China Normal University, China) &
Bin Xiong (East China Normal University, China)

Vol. 24 *Leningrad Mathematical Olympiads (1961–1991)*
by Dmitri Fomin

Vol. 23 *Solving Problems in Point Geometry:*
Insights and Strategies for Mathematical Olympiad and Competitions
by Jingzhong Zhang (Guangzhou University, China &
Chinese Academy of Sciences, China) &
Xicheng Peng (Central China Normal University, China)

Vol. 22 *Mathematical Olympiad in China (2021–2022):*
Problems and Solutions
editor-in-chief Bin Xiong (East China Normal University, China)

Vol. 21 *Problem Solving Methods and Strategies in High School*
Mathematical Competitions
edited by Bin Xiong (East China Normal University, China) &
Yijie He (East China Normal University, China)

The complete list of the published volumes in the series can be found at
http://www.worldscientific.com/series/mos

Vol. 29 | Mathematical Olympiad Series

Mathematical Olympiad

in China (2023)

Problems and Solutions

Editor-in-Chief

Xiong Bin

East China Normal University, China

Translators

Zhao Wei

East China Normal University, China

Chen Haoran

Xi'an Jiaotong-Liverpool University, China

Copy Editors

Shi Zhan
Zhang Liyu
Ni Ming

East China Normal University Press, China

East China Normal
University Press

World Scientific

Published by

East China Normal University Press
3663 North Zhongshan Road
Shanghai 200062
China

and

World Scientific Publishing Co. Pte. Ltd.
5 Toh Tuck Link, Singapore 596224
USA office: 27 Warren Street, Suite 401-402, Hackensack, NJ 07601
UK office: 57 Shelton Street, Covent Garden, London WC2H 9HE

Library of Congress Control Number: 2025005558

British Library Cataloguing-in-Publication Data
A catalogue record for this book is available from the British Library.

Mathematical Olympiad Series — Vol. 29
MATHEMATICAL OLYMPIAD IN CHINA (2023)
Problems and Solutions

Copyright © 2025 by East China Normal University Press and
World Scientific Publishing Co. Pte. Ltd.

ISBN 978-981-98-0848-9 (hardcover)
ISBN 978-981-98-0937-0 (paperback)
ISBN 978-981-98-0849-6 (ebook for institutions)
ISBN 978-981-98-0850-2 (ebook for individuals)

For any available supplementary material, please visit
https://www.worldscientific.com/worldscibooks/10.1142/14192#t=suppl

Desk Editors: Nambirajan Karuppiah/Angeline Husni

Typeset by Stallion Press
Email: enquiries@stallionpress.com

Preface

The first time China participate in IMO (International Mathematical Olympiad) was in 1985, when two students were sent to the 26th IMO. Since 1986, China has a team of 6 students at every IMO except in 1998. So far, up to 2024, China has achieved the number one ranking in team effort 24 times. A great majority of students received gold medals. The fact that China obtained such encouraging result is due to, on one hand, Chinese students' hard work and perseverance, and on the other hand, the effort of the teachers in schools and the training offered by national coaches. We believe this is also a result of the education system in China, in particular, the emphasis on training of the basic skills in science education.

The materials of this book come from two volumes (Vol. 2023) of a book series in Chinese "走向 IMO: 数学奥林匹克试题集锦" (*Forward to IMO:A Collection of Mathematical Olympiad Problems*). It is a collection of problems and solutions of the major mathematical competitions in China. It provides a glimpse of how the China national team is selected and formed. First, there is the China Mathematical Competition, a national event. It is held on the second Sunday of September every year. Through the competition, about 550 students are selected to join the China Mathematical Olympiad (commonly known as the winter camp), or in short CMO, in November. CMO lasts for five days. Both the type and the difficulty of the problems match those of IMO. Similarly, students are given three problems to solve in 4.5 hours each day. From CMO, 60 students are selected to forma national training team. The training takes place for two weeks in the month of March. After four to six tests plus two qualifying examinations, six students are finally selected to form the national team, taking part in IMO in July of that year.

In view of the differences in education, culture and economy of the western part of China with the coastal part in eastern China, mathematical competitions in West China did not develop as fast as the rest of the country. In order to promote the activity of mathematical competition, and to enhance the level of mathematical competition, starting from 2001, China Mathematical Olympiad Committee has been organizing the China Western Mathematical Olympiad.

Since 2012, the China Western Mathematical Olympiad has been renamed the China Western Mathematical Invitation. The competition dates have been changed from the first half of October to the middle of August since 2013.

The development of this competition reignited the enthusiasm of Western students for mathematics. Nowadays, the figure of Western students often appeared in the national team.

Since 1995, there had been no female students in the Chinese national team. In order to encourage more female students participating in the mathematical competition, starting from 2002, China Mathematical Olympiad Committee has been conducting the China Girls' Mathematical Olympiad. Again, the top 15 winners will be admitted directly into the CMO.

The authors of this book are coaches of the China national team. They are Xiong Bin, Xiao Liang, Wang Xinmao, Yao Yijun, Qu Zhenhua, Li Ting, Ai Yinhua, Wang Bin, Fu Yunhao, He Yijie and Zhang Sihui *et al.*

Those who took part in the translation work are Zhao Wei and Chen Haoran (improving solutions and providing helpful explanations to a few problems). We are grateful to Qiu Zonghu, Wang Changping and Liu Ruochuan for their guidance and assistance to the authors. We are grateful to Ni Ming, Shi Zhan and Zhang Liyu of East China Normal University Press. Their effort has helped make our job easier. We are also grateful to Tan Rok Ting and Angeline Husni of World Scientific Publishing for their hard work leading to the final publication of the book.

Introduction

Early Days

The International Mathematical Olympiad (IMO), founded in 1959, is one of the most competitive and highly intellectual activities in the world for high school students.

Even before IMO, there were already many countries which had mathematical competitions. They were mainly the countries in Eastern Europe and in Asia. In addition to the popularization of mathematics and the convergence in educational systems among different countries, the success of mathematical competitions at the national level provided a foundation for the setting-up of IMO. The countries that asserted great influence are Hungary, the former Soviet Union, and the United States. Here is a brief history of the IMO and mathematical competitions in China.

In 1894, the Department of Education in Hungary passed a motion and decided to conduct a mathematical competition for secondary schools. The well-known scientist, *J. von Eötvös*, was the Minister of Education at that time. His support for the event had made it a success and thus it was well publicized. In addition, the success of his son, *R. von Eötvös*, who was also a physicist, in proving the principle of equivalence of the general theory of relativity by *A. Einstein* through experiment, had brought Hungary to the world stage in science. Thereafter, the prize for mathematical competition in Hungary was named *"Eötvös* prize". This was the first formally organized mathematical competition in the world. In what follows, Hungary had indeed produced a lot of well-known scientists including *L. Fejér, G. Szegö, T. Radó, A. Haar* and *M. Riesz* (in real analysis), *D. König* (in combinatorics), *T. von Kármán* (in aerodynamics), and *J. C. Harsanyi* (in game theory), who had also won the Nobel Prize for Economics in 1994. They all were the winners of Hungary mathematical competition.

The top scientific genius of Hungary, *J. von Neumann*, was one of the leading mathematicians in the 20th century. *Neumann* was overseas while the competition took place. Later he did the competition himself and it took him half an hour to complete. Another mathematician worth mentioning is the highly productive number theorist *P. Erdös*. He was a pupil of *Fejér* and a winner of the Wolf Prize. *Erdös* was very passionate about mathematical competitions and setting competition questions. His contribution to discrete mathematics was unique and of great significance. The rapid progress and development of discrete mathematics over the subsequent decades had indirectly influenced the types of questions set in IMO. An internationally recognized prize was named after *Erdös* to honor those who had contributed to the education of mathematical competition. Professor *Qiu Zonghu* from China had won the prize in 1993.

In 1934, a famous mathematician *B. Delone* conducted a mathematical competition for high school students in Leningrad (now St. Petersburg). In 1935, Moscow also started organizing such events. Other than being interrupted during World War II, these events had been carried on until today. As for the Russian Mathematical Competition (later renamed as the Soviet Mathematical Competition), it was not started until 1961. Thus, the former Soviet Union and Russia became the leading powers of Mathematical Olympiad. A lot of grandmasters in mathematics, including the great *A. N. Kolmogorov*, were all very enthusiastic about the mathematical competition. They would personally involve themselves in setting the questions for the competition. The former Soviet Union even called it the Mathematical Olympiad, believing that mathematics is the "gymnastics of thinking". These points of view had a great impact on the educational community. The winner of the Fields Medal in 1998, *M. Kontsevich*, was once the first runner-up of the Russian Mathematical Competition. *G. Kasparov*, the international chess grandmaster, was once the second runner-up. *Grigori Perelman*, the winner of the Fields Medal in 2006 (which he declined), who solved the Poincaré's Conjecture, was a gold medalist of IMO in 1982.

In the United States of America, due to the active promotion by the renowned mathematician *G. D. Birkhoff* and his son, together with *G. Pólya*, the Putnam mathematics competition was organized in 1938 for junior undergraduates. Many of the questions were within the scope of high school curriculum. The top five contestants of the Putnam mathematical competition would be entitled to the membership of Putnam. Many of these eventually became outstanding mathematicians. There were the famous *R. Feynman* (winner of the Nobel Prize for Physics, 1965), *K. Wilson* (winner of the Nobel Prize for Physics, 1982), *J. Milnor* (winner of the Fields Medal, 1962), *D. Mumford* (winner of the Fields Medal, 1974), and *D. Quillen* (winner of the Fields Medal, 1978).

In 1972, in order to prepare for the IMO, the United States of America Mathematical Olympiad (USAMO) was established. The standard of questions posed was very high, parallel to that of the Winter Camp in China. Prior to this, the United States had organized American High School Mathematics Examination (AHSME) for the high school students since 1950. This was at the junior level and yet the most popular mathematics competition in America. Originally, it was intended to select about 100 contestants from AHSME to participate in USAMO. However, due to the discrepancy in the level of difficulty between the two competitions and other restrictions, from 1983 onwards, an intermediate level of competition, namely, American Invitational Mathematics Examination (AIME), was introduced. Henceforth both AHSME and AIME became internationally well-known. Since 2000, AHSME was replaced by AMC 12 and AMC 10. Students who perform well on the AMC 12 and AMC 10 are invited to participate in AIME. The combined scores of the AMC 12 and the AIME are used to determine approximately 270 individuals that will be invited back to take the USAMO, while the combined scores of the AMC 10 and the AIME are used to determine approximately 230 individuals that will be invited to take the USAJMO (United States of America Junior Mathematical Olympiad), which started in 2010 and follows the same format as the USAMO. A few cities in China had participated in the competition and the results have been encouraging.

Similar to the case of the former Soviet Union, the Mathematical Olympiad education was widely recognized in America. The book "How to Solve it" written by *George Polya* along with many other titles had been translated into many different languages. *George Polya* provided a whole series of general heuristics for solving problems of all kinds. His influence in the educational community in China should not be underestimated.

International Mathematical Olympiad

In 1956, the East European countries and the Soviet Union took the initiative to organize the IMO formally. The first International Mathematical Olympiad (IMO) was held in Brasov, Romania, in 1959. At that time, there were only seven participating countries, namely, Romania, Bulgaria,

Poland, Hungary, Czechoslovakia, East Germany and the Soviet Union. Subsequently, the United States of America, United Kingdom, France, Germany, and also other countries including those from Asia joined. Today, the IMO has managed to reach almost all the developed and developing countries. Except in the year 1980 due to financial difficulties faced by the host country, Mongolia, there have been 59 Olympiads held yearly and with 107 countries and regions participating nowadays.

The mathematical topics in the IMO include algebra, combinatorics, geometry, number theory. These areas have provided guidance for setting questions for the competitions. Other than the first few Olympiads, each IMO is normally held in mid-July every year, and the test paper consists of 6 questions in total. The actual competition lasts for 2 days for a total of 9 hours, where participants are required to complete 3 questions each day. Each question is 7 points, which totals up to 42 points. The full score for a team is 252 marks. About half of the participants will be awarded a medal, where 1/12 will be awarded a gold medal. The numbers of gold, silver and bronze medals awarded are in the ratio of 1:2:3, approximately. In the case when a participant provides a better solution than the official answer, a special award is given.

Each participating country and region will take turns to host the IMO. The cost is borne by the host country. China had successfully hosted the 31st IMO in Beijing. The event had made a great impact on the mathematical community in China. According to the rules and regulations of the IMO, all participating countries are required to send a delegation consisting of a leader, a deputy leader and 6 contestants. The problems are contributed by the participating countries and are later selected carefully by the host country for submission to the international jury set up by the host country. Eventually, only 6 problems will be accepted for use in the competition. The host country does not provide any questions. The shortlisted problems are subsequently translated, if necessary, in English, French, German, Spanish, Russian, and other working languages. After that, the team leaders will translate the problems into their own languages.

The answer scripts of each participating team will be marked by the team leader and the deputy leader. The team leader will later present the scripts of their contestants to the coordinators for assessment. If there is any dispute, the matter will be settled by the jury. The jury is formed by the various team leaders and an appointed chairman by the host country. The jury is responsible for deciding the final 6 problems for the competition. Their duties also include finalizing the grading standard, ensuring the accuracy of the translation of the problems, standardizing replies to written queries raised by participants during the competition, synchronizing differences in grading between the team leaders and the coordinators, and also deciding on the cut-off points for the medals depending on the contestants' results as the difficulties of problems each year are different.

China had participated informally in the 26th IMO in 1985. Only two students were sent. Starting from 1986, except in 1998 when the IMO was held in Taiwan, China had always sent 6 official contestants to the IMO. Today, the Chinese contestants not only performed outstandingly in the IMO, but also in the International Physics, Chemistry, Informatics, and Biology Olympiads. This can be regarded as an indication that China pays great attention to the training of basic skills in mathematics and science education.

Winners of the IMO

Among all the IMO medalists, there were many of them who eventually became great mathematicians. They were also awarded the Fields Medal, Wolf Prize and Nevanlinna Prize (a prominent mathematics prize for computing and informatics). In what follows, we name some of the winners.

G. Margulis, a silver medalist of IMO in 1959, was awarded the Fields Medal in 1978. *L. Lovasz*, who won the Wolf Prize in 1999, was awarded the Special Award in IMO consecutively in 1965 and 1966. *V. Drinfeld*, a gold medalist of IMO in 1969, was awarded the Fields Medal in 1990. *J.-C. Yoccoz* and *T. Gowers*, who were both awarded the Fields Medal in 1998, were gold medalists in IMO in 1974 and 1981, respectively. A silver medalist of IMO in 1985, *L. Lafforgue*, won the Fields Medal in 2002. A gold medalist of IMO in 1982, *Grigori Perelman* from Russia, was awarded the Fields Medal in 2006 for solving the final step of the Poincaré conjecture. In 1986, 1987, and 1988, *Terence Tao* won a bronze, silver, and gold medal respectively. He was the youngest participant to date in the IMO, first competing at the age of ten. He was also awarded the Fields Medal in 2006. Gold medalist of IMO 1988 and 1989, *Ngo Bau Chao*, won the Fields Medal in 2010, together with the bronze medalist of IMO 1988, *E.Lindenstrauss*. Gold medalist of IMO 1994 and 1995, *Maryam Mirzakhani* won the Fields Medal in 2014. A gold medalist of IMO in 1995, Artur Avila, won the Fields Medal in 2014. Gold medalist of IMO 2005, 2006 and 2007, Peter Scholze, won the Fields Medal in 2018. A Bronze medalist of IMO in 1994, Akshay Venkatesh, won the Fields Medal in 2018.

A silver medalist of IMO in 1977, *P. Shor*, was awarded the Nevanlinna Prize. A gold medalist of IMO in 1979, *A. Razborov*, was awarded the Nevanlinna Prize. Another gold medalist of IMO in 1986, *S. Smirnov*, was awarded the Clay Research Award. *V. Lafforgue*, a gold medalist of IMO in 1990, was awarded the European Mathematical Society prize. He is *L. Lafforgue*'s younger brother.

Also, a famous mathematician in number theory, *N. Elkies*, who is also a professor at Harvard University, was awarded a gold medal of IMO in 1982. Other winners include *P. Kronheimer*, awarded a silver medal in 1981, and *R. Taylor* a contestant of IMO in 1980.

Mathematical Competition in China

Due to various reasons, mathematical competition in China started relatively late but is progressing vigorously.

"We are going to have our own mathematical competition too!" said *Hua Luogeng*. *Hua* is a household name in China. The first mathematical competition was held concurrently in Beijing, Tianjin, Shanghai, and Wuhan in 1956. Due to the political situation at the time, this event was interrupted a few times. It was not until 1962, when the political environment started to improve, that Beijing and other cities started organizing the competition, though not regularly. In the era of Cultural Revolution, the whole educational system in China was in chaos. The mathematical competition came to a complete halt. In contrast, the mathematical competition in the former Soviet Union was still on-going during the war and at a time of difficult political situation. The competitions in Moscow were interrupted only 3 times between 1942 and 1944. It was indeed commendable.

In 1978, it was the spring of science. *Hua Luogeng* conducted the Middle School Mathematical Competition for 8 provinces in China. The mathematical competition in China was then making a fresh start and embarked on a road of rapid development. *Hua* passed away in 1985. To commemorate him, a competition named *Hua Luogeng* Gold Cup was set up in 1986 for students in Grades 6 and 7, and it has had a great impact.

The mathematical competitions in China before 1980 can be considered as the initial period. The problem sets were within the scope of middle school textbooks. After 1980, the competitions gradually moved toward the senior middle school level. In 1981, the Chinese Mathematical Society decided to conduct the China Mathematical Competition, a national event for high schools.

In 1981, the United States of America, the host country of IMO, issued an invitation to China to participate in the event. Only in 1985, China sent two contestants to participate informally in the IMO. The results were not encouraging. In view of this, another activity called the Winter Camp was conducted after the China Mathematical Competition. The Winter Camp was later renamed as the China Mathematical Olympiad or CMO. The winning team would be awarded the *Chern Shiing-Shen* Cup. Based on the outcome at the Winter Camp, a selection would be made to form the 6-member national team for IMO. From 1986 onwards, other than the year when IMO was organized in Taiwan, China has been sending a 6-member team to IMO. Up to 2018, China had been awarded the overall team champion for 19 times.

In 1990, China successfully hosted the 31st IMO. It showed that the standard of mathematical competition in China has leveled that of other leading countries. Firstly, the fact that China achieves the highest marks at the 31st IMO for the team is evidence for the effectiveness of the pyramid approach in selecting the contestants in China. Secondly, the Chinese mathematicians had simplified and modified over 100 problems and submitted them to the team leaders of the 35 countries for their perusal. Eventually, 28 problems were recommended. At the end, 5 problems were chosen (IMO requires 6 problems). This is also evidence to show that China has achieved the highest quality in setting problems. Thirdly, the answer scripts of the participants were marked by the various team leaders and assessed by the coordinators who were nominated by the host countries. China had formed a group 50 mathematicians to serve as coordinators who would ensure the high accuracy and fairness in marking. The marking process was completed half a day earlier than it was scheduled. Fourthly, that was the first ever IMO organized in Asia. The outstanding performance by China had encouraged the other developing countries, especially those in Asia. The organizing and coordinating work of the IMO by the host country was also reasonably good.

In China, the outstanding performance in mathematical competition is a result of many contributions from all quarters of the mathematical community. There are the older generation of mathematicians, middle aged mathematicians, and also the middle and elementary school teachers. There is one person who deserves a special mention, and he is *Hua Luogeng*. He initiated and promoted mathematical competitions. He is also the author of the following books: Beyond *Yang hui*'s Triangle, Beyond the *pi* of *Zu Chongzhi*, Beyond the Magic Computation of *Sun-zi*, Mathematical Induction, and

Mathematical Problems of Bee Hive. These were his books derived from mathematical competitions. When China resumed mathematical competitions in 1978, he participated in setting problems and giving critique to solutions of the problems. Other outstanding books derived from the Chinese mathematical competitions are: Symmetry by *Duan Xuefu*, Lattice and Area by *Min Sihe*, One Stroke Drawing and Postman Problem by *Jiang Boju*.

After 1980, the younger mathematicians in China had taken over from the older generation of mathematicians in running the mathematical competitions. They worked and strived hard to bring the level of mathematical competition in China to a new height. *Qiu Zonghu* is one such outstanding representative. From the training of contestants and leading the team 3 times to IMO, to the organizing of the 31th IMO in China, he had contributed prominently and was awarded the *P. Erdös* prize.

Preparation for IMO

Currently, the selection process of participants for IMO in China is as follows.

First, the China Mathematical Competition, a national competition for high schools, is organized on the second Sunday in September every year. The objectives are to increase the interest of students in learning mathematics, to promote the development of co-curricular activities in mathematics, to help improve the teaching of mathematics in high schools, to discover and cultivate the talents, and also to prepare for the IMO. This has been happening since 1981. Currently, there are about 500,000 participants taking part in it.

Through the China Mathematical Competition, around 550 students are selected to take part in the China Mathematical Olympiad or CMO, that is, the Winter Camp. The CMO lasts for 5 days and is held in November every year. The types and difficulties of the problems in CMO are very much similar to the IMO. There are also 3 problems to be completed within 4.5 hours each day. However, the score for each problem is 21 marks which adds up to 126 marks in total. Starting from 1990, the Winter Camp instituted the *Chern Shiing-Shen* Cup for team championship. In 1991, the Winter Camp was officially renamed as the China Mathematical Olympiad (CMO). It is similar to the highest national mathematical competition in the former Soviet Union and the United States.

The CMO awards the first, second and third prizes. Among the participants of CMO, about 60 students are selected to participate in the training
for IMO. The training takes place in March every year. After 6 to 8 tests
and another 2 rounds of qualifying examinations, only 6 contestants are
short-listed to form the China IMO national team to take part in the IMO
in July.

Besides the China Mathematical Competition (for high schools), the
Junior Middle School Mathematical Competition is also developing well.
Starting from 1984, the competition is organized in April every year by
the Popularization Committee of the Chinese Mathematical Society. The
various provinces, cities and autonomous regions would rotate to host the
event. Another mathematical competition for the junior middle schools is
also conducted in April every year by the Middle School Mathematics Education Society of the Chinese Educational Society since 1998 till 2014.

The *Hua Luogeng* Gold Cup, a competition by invitation, has also
been successfully conducted since 1986. The participating students comprise
elementary-six and junior-middle-one students. The format of the competition consists of a preliminary round, semi-finals in various provinces, cities
and autonomous regions, then the finals.

Mathematical competitions in China provide a platform for students
to showcase their talents in mathematics. It encourages learning of mathematics among students. It helps identify talented students and to provide
them with differentiated learning opportunities. It develops co-curricular
activities in mathematics. Finally, it brings about changes in the teaching
of mathematics.

Contents

1

China Mathematical Competition (First Round)

2022

While the scope of the test questions in the first round of the 2022 China Mathematical Competition does not exceed the teaching requirements and content specified in the 'General High School Mathematics Curriculum Standards (2017)' promulgated by the Ministry of Education of China in 2017, the methods of proposing the questions have been improved. The emphasis placed on to test the students' basic knowledge and skills, and their abilities to integrate and use flexibly of them. Each test paper includes 8 fill-in-the-blank questions and 3 answer questions. The answer time is 80 minutes, and the full score is 120 marks.

The scope of the test questions in the Second Round (Complementary Test) is in line with the International Mathematical Olympiad, with some expanded knowledge, plus a few contents of the Mathematical Competition Syllabus. Each test paper consists of 4 answer questions, including a plane geometry one, and the answering time is 170 minutes. The full score is 180 marks.

Test Paper A

Part I Short-Answer Questions (Questions 1–8, eight marks each)

1 The sum of all elements in set $A = \{n|n^3 < 2022 < 3^n, \in \mathbf{Z}\}$ is

_____.

Solution When n is an integer, $n^3 < 2022 \Leftrightarrow n \leqslant 12$, $3^n > 2022 \Leftrightarrow n \geqslant 7$.
Therefore, $A = \{7, 8, 9, 10, 11, 12\}$. The required sum is $7 + 8 + \cdots + 12 = 57$.

2 Suppose function $f(x) = \frac{x^2+x+16}{x}$ $(2 \leqslant x \leqslant a)$, where the real number a is greater than 2. If the range of $f(x)$ is $[9, 11]$, then the range of a is _____.

Solution Consider function $f_0(x) = x + \frac{16}{x} + 1$ $(x \geqslant 2)$. (When $2 \leqslant x \leqslant a$, $f(x) = f_0(x)$).
Since $f_0(x)$ is strictly decreasing on $[2, 4]$ and strictly increasing on $[4, +\infty)$ and noting that $f_0(2) = 11$, $f_0(4) = 9$, $f_0(8) = 11$, the range of desired a is $a \in [4, 8]$.

3 An uneven coin is tossed twice. If the probability of getting heads in both random tosses is 9 times the probability of getting tails in both tosses, then the probability of getting one head and one tail is _____.

Solution Suppose the coin is tossed once randomly and the probability of getting heads is $p(0 \leqslant p \leqslant 1)$, and the probability of getting tails is $1 - p$. According to the question, we have $p^2 = 9(1 - p)^2$, and thus $p = \frac{3}{4}$.
Hence, the probability of randomly tossing it twice and getting one head and one tail is $2p(1 - p) = \frac{3}{8}$.

4 If complex number z satisfies: $\frac{z-3i}{z+i}$ is a negative real number (i is the imaginary unit), and $\frac{z-3}{z+1}$ is a pure imaginary number, then the value of z is _____.

Solution Let $z = a + bi(a, b \in \mathbf{R})$.
Since $\frac{z-3i}{z+i} = \frac{a+(b-3)i}{a+(b+1)i}$ is a negative real number, that is equivalent to

$$(a + (b - 3)i)(a - (b + 1)i) < 0.$$

Hence, $a^2 + (b-3)(b+1) < 0$ and $-4a = 0$, namely, $a = 0$ and $b-3$ and $b+1$ have opposite signs.

And since $\frac{z-3}{z+1} = \frac{-3+bi}{1+bi}$ is a pure imaginary number, $\text{Re}((-3+bi)(1-bi)) = b^2 - 3 = 0$. Combined with the fact that $b-3$ and $b+1$ have opposite signs, we can get $b = \sqrt{3}$. Therefore, $z = \sqrt{3}i$.

5 Given pyramid $P - ABCD$, the edge lengths of AB and BC are $\sqrt{2}$, and all other edges have lengths of 1. Then the volume of this pyramid is _____.

Solution Let O be the projection of P onto the base $ABCD$, and $PO = h$.

Since $PA = PB = PC = PD = 1$, it follows that $OA = OB = OC = OD = \sqrt{1-h^2}$. Thus, $ABCD$ is a cyclic quadrilateral.

Furthermore, since $AB = BC = \sqrt{2}, AD = CD = 1$, $ABCD$ is a kite with BD as its axis of symmetry. Hence, $\angle BAD = \angle BCD = 90°$, and O is the midpoint of BD. So $\sqrt{1-h^2} = \frac{BD}{2} = \frac{\sqrt{3}}{2}$, we solve for h and get $h = \frac{1}{2}$.

Therefore, the volume of this pyramid is $\frac{1}{3}S_{ABCD} \cdot h = \frac{1}{3}AB \cdot AD \cdot h = \frac{\sqrt{2}}{6}$.

6 Given that the graph of function $y = f(x)$ is symmetric about point $(1,1)$ and also symmetric about line $x + y = 0$. If $x \in (0,1)$ and $f(x) = \log_2(x+1)$, then the value of $f(\log_2 10)$ is _____.

Solution Let Γ represent the graph of function $y = f(x)$.

For any $x_0 \in (0,1)$, let $y_0 = \log_2 (1+x_0)$. Hence, $(x_0, y_0) \in \Gamma$ and $y_0 \in (0,1)$.

Using the central symmetry and axial symmetry of Γ, it can be deduced in turn that

$$(2-x_0, 2-y_0) \in \Gamma, \quad (y_0 - 2, x_0 - 2) \in \Gamma, \quad (4-y_0, 4-x_0) \in \Gamma.$$

Take $x_0 = \frac{3}{5}$, and then $4 - y_0 = 4 - \log_2(1+x_0) = \log_2 10$.
Therefore, $f(\log_2 10) = f(4 - y_0) = 4 - x_0 = 4 - \frac{3}{5} = \frac{17}{5}$.

7 In a plane rectangular coordinate system, consider the ellipse Ω : $\frac{x^2}{4} + y^2 = 1$. Let P be a moving point on Ω, and A, B are two fixed points with B at $(0,3)$. If the area of $\triangle PAB$ has a minimum value of 1 and a maximum value of 5, then the length of segment AB is _____.

Solution　It is obvious that line AB cannot intersect the ellipse. Let the equation of AB be $y = kx + 3$.

Let the moving point P be at $P(2\cos\theta, \sin\theta)$, and then the distance from P to line AB is $d(\theta) = \frac{|2k\cos\theta - \sin\theta + 3|}{\sqrt{k^2+1}}$.

Note that the length of segment AB is fixed. According to the question, when θ changes, the ratio of the maximum value of $d(\theta) = \frac{2S_{\triangle PAB}}{|AB|}$ to its minimum value is 5. Specifically, $2k\cos\theta - \sin\theta + 3$ cannot be 0, so its value is always positive or always negative.

Since the maximum value of $2k\cos\theta - \sin\theta + 3$ is $3 + \sqrt{4k^2 + 1}$ which is positive, its minimum value $3 - \sqrt{4k^2 + 1}$ is also positive. Hence, $\frac{3+\sqrt{4k^2+1}}{3-\sqrt{4k^2+1}} = 5$. Solving this gives $k^2 = \frac{3}{4}$. Thus, the minimum value of $d(\theta)$ is $d_0 = \frac{3-\sqrt{4k^2+1}}{\sqrt{k^2+1}} = \frac{2}{\sqrt{7}}$.

Since the minimum value of $S_{\triangle PAB}$ is 1, we have $\frac{1}{2}|AB| \cdot d_0 = 1$. Therefore, $|AB| = \frac{2}{d_0} = \sqrt{7}$.

8　In a unit square, if two sides are colored in color i and the other two sides are colored in different colors other than i, this unit square is said to be 'i-dominant'. As shown in Fig. 1.1, a 1×3 grid table with a total of 10 unit-length segments, which are colored in colors: red, yellow, and blue. Each segment is colored in one of these three colors such that there is one red-dominant, one yellow-dominant, and one blue-dominant unit square, respectively. The number of such coloring schemes is _____. (Express your answer as a numerical value.)

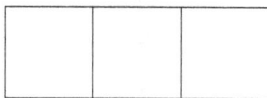

Fig. 1.1

Solution　Let the unit square be referred to as 'a square'. The following conclusions are apparent for the coloring method that fits the question:

(1) When one side of a square is already colored in a certain color i, there are exactly $3! = 6$ ways of coloring the other three sides if the square is i-dominant; there are exactly 3 ways of coloring the other three sides if the square is $j(\neq i)$-dominant.

(2) When a square already has two sides colored in a certain color i, the square can only be i-dominant, and there are exactly 2 ways to color the other two sides.

(3) When a square already has two sides colored in some two colors i, j, if the square is either i-dominant or j-dominant, then there are exactly two ways of coloring the other two sides; if the square is $k(\neq i, j)$-dominant, then there is a unique way of coloring the other two sides.

Considering symmetry, we only need to calculate the number N of coloring ways for the left, middle, and right squares being red-dominant, yellow-dominant, and blue-dominant, respectively. Hence, the result of this question is $6N$. If the common side of the left and middle squares is colored in color i, and the common side of the middle and right squares is colored in color j, then this coloring scheme is classified as the "i-j" type.

Using the above conclusions (1), (2), and (3), we can find the number of such coloring schemes as shown in the following table:

Red–Red: 0	Red–Yellow: $6 \times 2 \times 3 = 36$	Red–Blue: $6 \times 1 \times 6 = 36$
Yellow–Red: $3 \times 2 \times 3 = 18$	Yellow–Yellow: $3 \times 2 \times 3 = 18$	Yellow–Blue: $3 \times 2 \times 6 = 36$
Blue–Red: $3 \times 1 \times 3 = 9$	Blue–Yellow: $3 \times 2 \times 3 = 18$	Blue–Blue: 0

Thus, $N = 3 \times 36 + 3 \times 18 + 9 = 171$. Therefore, the final result is $6N = 1026$.

Part II Word Problems (16 marks for Question 9, 20 marks for Question 10 and 11, and then 56 marks in total)

9 (16 marks) Given that the interior angles A, B and C of $\triangle ABC$ satisfy $\sin A = \cos B = \tan C$, find the value of $\cos^3 A + \cos^2 A - \cos A$.

Solution From $\sin A = \cos B$, we know that $A = \frac{\pi}{2} \pm B$. However, since $\tan C$ is finite, C cannot be a right angle. Thus, it can only be $A = \frac{\pi}{2} + B$, which in turn gives $C = \pi - A - B = \frac{3\pi}{2} - 2A$.

Therefore, $\sin A = \tan C = \tan\left(\frac{3\pi}{2} - 2A\right) = \cot 2A$, and hence

$$1 = \sin A \cdot \tan 2A = \sin A \cdot \frac{\sin 2A}{\cos 2A}$$

$$= \sin A \cdot \frac{2\sin A \cos A}{2\cos^2 A - 1}$$

$$= \frac{2(1 - \cos^2 A) \cdot \cos A}{2\cos^2 A - 1}.$$

The above equation is equivalent to

$$2\cos^2 A - 1 = 2(1 - \cos^2 A) \cdot \cos A.$$

Thus,

$$\cos^3 A + \cos^2 A - \cos A = \frac{1}{2}.$$

10 **(20 marks)** Given a positive integer $m(m \geqslant 3)$. Consider positive arithmetic sequence $\{a_n\}$ and positive geometric sequence $\{b_n\}$ such that the first term of $\{a_n\}$ equals the common ratio of $\{b_n\}$, the first term of $\{b_n\}$ equals the common difference of $\{a_n\}$, and $a_m = b_m$. Find the minimum value of a_m, and determine the ratio of a_1 to b_1 when a_m takes this minimum value.

Solution Based on the question, let the first term of $\{a_n\}$ be a and the common difference be b, and let the first term of $\{b_n\}$ be b and the common ratio be a, where $a, b > 0$. Then $a_m = a + (m-1)b, b_m = b \cdot a^{m-1}$.

Denote $\lambda = a_m = b_m$. Then from the above expressions, we have

$$\lambda = b \cdot a^{m-1} = b \cdot (\lambda - (m-1)b)^{m-1}.$$

Using the AM-GM inequality for m elements, we get

$$(m-1)^2\lambda = (m-1)^2 b \cdot (\lambda - (m-1)b)^{m-1}$$
$$\leqslant \left(\frac{(m-1)^2 b + (m-1)(\lambda - (m-1)b)}{m}\right)^m = \left(\frac{(m-1)\lambda}{m}\right)^m,$$

which leads to $\lambda \geqslant \left(\frac{m^m}{(m-1)^{m-2}}\right)^{\frac{1}{m-1}}$.

The equality holds when $(m-1)^2 b = \lambda - (m-1)b$, i.e., $b = \frac{\lambda}{m(m-1)}$.

At this point λ (i.e., a_m) takes its minimum value $\left(\frac{m^m}{(m-1)^{m-2}}\right)^{\frac{1}{m-1}}$, and accordingly we have $a = \lambda - (m-1)b = (m-1)^2 b$. Therefore,

$$\frac{a_1}{b_1} = \frac{a}{b} = (m-1)^2.$$

11 **(20 marks)** Consider hyperbola $\Gamma : \frac{x^2}{3} - y^2 = 1$ in a plane rectangular coordinate system. For any point P in the plane that is not on Γ, denote Ω_P as the set of all lines passing through point P and intersecting Γ at two points. For any line $l \in \Omega_P$, let M and N be the two intersection points of l and Γ, and define $f_P(l) = |PM| \cdot |PN|$. If there exists a line $l_0 \in \Omega_P$ such that l_0 intersects Γ at two points on opposite sides of the y-axis, and for any other line $l \in \Omega_P, l \neq l_0$, there is $f_P(l) > f_P(l_0)$, then P is called a 'good point'. Find the area of the region formed by all good points.

Solution　Let $P(a, b)$ be a good point. Consider the necessary and sufficient conditions for a and b.

For any line $l \in \Omega_P$, let the angle of inclination of l be $\theta(0 \leqslant \theta < \pi)$. The parametric equation of l can be written as

$$l : \begin{cases} x = a + t\cos\theta, \\ y = b + t\sin\theta \end{cases} \quad (t \text{ is the parameter}). \qquad \textcircled{1}$$

Substituting $\textcircled{1}$ into the equation of Γ yields the equation in t:

$$(3\sin^2\theta - \cos^2\theta)t^2 + 2(3b\sin\theta - a\cos\theta)t + 3b^2 + 3 - a^2 = 0 \qquad \textcircled{2}$$

According to the question, $\textcircled{2}$ has two distinct real solutions t_1 and t_2, which is equivalent to $3\sin^2\theta - \cos^2\theta \neq 0$, and the discriminant

$$\triangle = 4(3b\sin\theta - a\cos\theta)^2 - 4(3\sin^2\theta - \cos^2\theta)(3b^2 + 3 - a^2) > 0.$$

Simplifying this, we get

$$\begin{cases} \theta \neq \dfrac{\pi}{6}, \dfrac{5\pi}{6}, \\ (a\sin\theta - b\cos\theta)^2 > 3\sin^2\theta - \cos^2\theta. \end{cases} \qquad \textcircled{3}$$

When the angle of inclination θ satisfies $\textcircled{3}$, from the geometric meaning of the parameters t in $\textcircled{1}$ and the definition of $f_P(l)$, we have

$$f_P(l) = |t_1| \cdot |t_2| = \frac{|3b^2 + 3 - a^2|}{|3\sin^2\theta - \cos^2\theta|},$$

where $|3b^2 + 3 - a^2| > 0$ because P is not on Γ.

When l intersects Γ at two points on opposite sides of the y-axis, from the properties of hyperbola, we know $\theta \in \left(0, \frac{\pi}{6}\right) \cup \left(\frac{5\pi}{6}, \pi\right)$.

At this point, $3\sin^2\theta - \cos^2\theta < 0$. It is obvious that θ satisfies $\textcircled{3}$, and $f_P(l) = \frac{|3b^2 + 3 - a^2|}{\cos^2\theta - 3\sin^2\theta}$. $f_P(l)$ takes the minimum value $f_P(l_0) = |3b^2 + 3 - a^2|$ if and only if $\theta = 0$, where line $l_0 : y = b$.

When l intersects Γ at two points on the same side of the y-axis, it is required that θ satisfies $\textcircled{3}$ and $\theta \in \left(\frac{\pi}{6}, \frac{5\pi}{6}\right)$. According to the problem, for any such l (if it exists), there is $f_P(l) = \frac{|3b^2 + 3 - a^2|}{3\sin^2\theta - \cos^2\theta} > f_P(l_0)$, that is, $0 < 3\sin^2\theta - \cos^2\theta < 1$. This is equivalent to $\theta \in \left(\frac{\pi}{6}, \frac{\pi}{4}\right) \cup \left(\frac{3\pi}{4}, \frac{5\pi}{6}\right)$.

In other words, a and b need to satisfy: for any $\theta \in \left[\frac{\pi}{4}, \frac{3\pi}{4}\right]$, ③ does not hold, i.e., for any $\theta \in \left[\frac{\pi}{4}, \frac{3\pi}{4}\right]$, there is

$$|a\sin\theta - b\cos\theta| \leqslant \sqrt{3\sin^2\theta - \cos^2\theta}. \qquad ④$$

In ④, let $\theta = \frac{\pi}{4}, \frac{3\pi}{4}$, we get $\left|\frac{\sqrt{2}}{2}a - \frac{\sqrt{2}}{2}b\right| \leqslant 1$, $\left|\frac{\sqrt{2}}{2}a + \frac{\sqrt{2}}{2}b\right| \leqslant 1$, respectively. Therefore,

$$|a| + |b| \leqslant \sqrt{2}.$$

Conversely, when $|a| + |b| \leqslant \sqrt{2}$, note that when $\theta a \in \left[\frac{\pi}{4}, \frac{3\pi}{4}\right]$, there is $\sin\theta \geqslant |\cos\theta| \geqslant 0$.

Thus,

$$
\begin{aligned}
|a\sin\theta - b\cos\theta| &\leqslant |a|\sin\theta + |b||\cos\theta| \\
&\leqslant \left(\sqrt{2} - |b|\right)\sin\theta + |b||\cos\theta| \\
&\leqslant \sqrt{2}\sin\theta \leqslant \sqrt{2\sin^2\theta + (\sin^2\theta - \cos^2\theta)} \\
&= \sqrt{3\sin^2\theta - \cos^2\theta},
\end{aligned}
$$

namely, ④ holds.

Therefore, $P(a, b)$ is a good point if and only if $|a| + |b| \leqslant \sqrt{2}$. So the region of all good points is $A = \{(a, b)||a| + |b| \leqslant \sqrt{2}\}$. The desired area is $S_A = 4$.

Test Paper B

Part I Short-Answer Questions (Questions 1–8, eight marks each)

1   The solution set of the inequality $\frac{20}{x-9} > \frac{22}{x-11}$ is _____.

Solution   By transposing and combining the terms, we get

$$\frac{20(x-11) - 22(x-9)}{(x-9)(x-11)} > 0,$$

which is equivalent to $\frac{x+11}{(x-9)(x-11)} < 0$. It is easy to see that the solution set is $(-\infty, -11) \cup (9, 11)$.

2   In a plane rectangular coordinate system, circle Ω is drawn with its center at the focus of parabola $\Gamma : y^2 = 6x$ and is tangent to the directrix of Γ. Then the area of circle Ω is _____.

Solution   The distance from the focus of parabola Γ to its directrix is 3, so the radius of circle Ω is $r = 3$.

Therefore, the area of circle Ω is $\pi r^2 = 9\pi$.

3 The maximum value of function $f(x) = \log_{10} 2 \cdot \log_{10} 5 - \log_{10} 2x \cdot \log_{10} 5x$ is _____.

Solution

$$f(x) = \log_{10} 2 \cdot \log_{10} 5 - (\log_{10} 2 + \log_{10} x) \cdot (\log_{10} 5 + \log_{10} x)$$

$$= -(\log_{10} 2 + \log_{10} 5) \log_{10} x - \log_{10}^2 x$$

$$= -\log_{10} x - \log_{10}^2 x = -\left(\log_{10} x + \frac{1}{2}\right)^2 + \frac{1}{4} \leqslant \frac{1}{4}.$$

When $\log_{10} x = -\frac{1}{2}$, that is, $x = \frac{\sqrt{10}}{10}$, $f(x)$ takes the maximum value $\frac{1}{4}$.

4 An uneven coin is tossed twice randomly. If the probability of getting heads twice in two tosses is $\frac{1}{2}$, then the probability of getting one head and one tail in two tosses is _____.

Solution   Let $p(0 \leqslant p \leqslant 1)$ be the probability of getting head in a single toss $(0 \leqslant p \leqslant 1)$. According to the question, $p^2 = \frac{1}{2}$, so $p = \frac{\sqrt{2}}{2}$. Therefore, the probability of getting one head and one tail in two random tosses is $2p(1 - p) = \sqrt{2} - 1$.

5 Given that complex number z satisfies $|z| = 1$ and $\operatorname{Re}\frac{z+1}{\bar{z}+1} = \frac{1}{3}$, then the value of $\operatorname{Re}\frac{z}{\bar{z}}$ is _____.

Solution   Since $z\bar{z} = |z|^2 = 1$, we have $\bar{z} = \frac{1}{z}$. Therefore,

$$\operatorname{Re}\frac{z+1}{\bar{z}+1} = \operatorname{Re}\frac{z+1}{\frac{1}{z}+1} = \operatorname{Re} z = \frac{1}{3}.$$

Given that $|z| = 1$, we can get $z = \frac{1}{3} \pm \frac{2\sqrt{2}}{3}\mathrm{i}$.

Hence, $\operatorname{Re}\frac{z}{\bar{z}} = \operatorname{Re} z^2 = \operatorname{Re}\left(-\frac{7}{9} \pm \frac{4\sqrt{2}}{9}\mathrm{i}\right) = -\frac{7}{9}$.

6 If the edges of right square pyramid $P - ABCD$ are all of equal length, and M is the midpoint of edge AB, then the cosine of the angle between skew lines BP and CM is _____.

Solution   Take the midpoint of AP as N, and then $MN \parallel BP$. So the angle between skew lines BP and CM is $\angle CMN$ (or its supplement angle).

Without loss of generality, let the edge lengths of right square pyramid $P - ABCD$ be 2. It is easy to see that $AC = 2\sqrt{2}$, and thus $\angle APC = 90°$. Therefore, $CN = \sqrt{PC^2 + PN^2} = \sqrt{5}$.

And since $MN = 1, CM = \sqrt{5} = CN$, we can get

$$\cos\angle CMN = \frac{MN}{2CM} = \frac{\sqrt{5}}{10},$$

that is, the desired value is $\frac{\sqrt{5}}{10}$.

7   If the three interior angles A, B, C of $\triangle ABC$ satisfy $\cos A = \sin B = 2\tan\frac{C}{2}$, then the value of $\sin A + \cos A + 2\tan A$ is _____.

Solution   Since $\cos A = \sin B$, we know $B = \frac{\pi}{2} \pm A$.

Suppose $B = \frac{\pi}{2} - A$, and then $C = \frac{\pi}{2}$. In this case, $\cos A = \sin B = 2$, which is a contradiction.

Thus, there can only be $B = \frac{\pi}{2} + A$, and then $C = \frac{\pi}{2} - 2A$. Therefore,

$$\cos A = 2\tan\frac{C}{2} = 2\tan\left(\frac{\pi}{4} - A\right) = 2 \cdot \frac{1 - \tan A}{1 + \tan A},$$

and this is equivalent to $\cos A \cdot (1 + \tan A) = 2 - 2\tan A$.

Consequently, $\sin A + \cos A + 2\tan A = 2$.

8   Among the four sides of a unit square, if three sides are colored in three different colors, it is called 'multicolored'. As shown in Fig. 1.2, a 1×3 grid has a total of 10 unit-length segments. We need to color these 10 segments, each with one of the three colors: red, yellow, or blue, in such a way that all three unit squares are multicolored. The number of such coloring schemes is _____. (Express your answer as a numerical value)

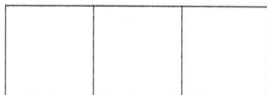

Fig. 1.2

Solution   We say 'square' short for the unit square.

Lemma   *If one side of a square has already been colored, there are exactly 12 ways to color the remaining three sides to make the square multicolored.*

In fact, assuming the already colored side is red, the remaining three sides can be colored in permutations of red, yellow, and blue, or two yellows and one blue, or two blues and one yellow. There are $3! + 3 + 3 = 12$ ways in total.

First, color the common side of the left and middle squares, which has 3 ways.

Then complete the coloring of the left square. According to the lemma, there are 12 ways.

Similarly, there are 12 ways to complete the coloring of the middle square.

At this point, the right square has one side already colored, so there are 12 ways to complete its coloring.

By the multiplication principle, the number of coloring schemes satisfying the problem is $3 \times 12 \times 12 \times 12 = 5184$.

Part II Word Problems (16 marks for Question 9, 20 marks for Questions 10 and 11, and then 56 marks in total)

9 **(16 marks)** In a plane rectangular coordinate system, F_1 and F_2 are the foci of hyperbola $\Gamma : \frac{x^2}{3} - y^2 = 1$. Point P on Γ satisfies $\overrightarrow{PF_1} \cdot \overrightarrow{PF_2} = 1$. Find the sum of the distances from P to the two asymptotes of Γ.

Solution It is easy to know that the coordinates of F_1, F_2 are $(\pm 2, 0)$, and the equations of the asymptotes of Γ are $x + \sqrt{3}y = 0$ and $x - \sqrt{3}y = 0$, respectively.

Let $P(u, v)$ and denote the sum of the distances from P to the two asymptotes by S. Then $S = \frac{|u+\sqrt{3}v|}{2} + \frac{|u-\sqrt{3}v|}{2}$.

Since $\overrightarrow{PF_1} \cdot \overrightarrow{PF_2} = (u-2)(u+2) + v^2 = u^2 + v^2 - 4$, we have $u^2 + v^2 - 4 = 1$, that is, $u^2 + v^2 = 5$.

And since P is on Γ, we have $\frac{u^2}{3} - v^2 = 1$. Solving this gives $u^2 = \frac{9}{2}, v^2 = \frac{1}{2}$. Note that $|u| > \sqrt{3}|v|$, so $S = \frac{2|u|}{2} = |u| = \frac{3\sqrt{2}}{2}$. Therefore, the desired sum of distances is $\frac{3\sqrt{2}}{2}$.

10 **(20 marks)** Suppose positive numbers a_1, a_2, a_3, b_1, b_2 and b_3 satisfy: a_1, a_2 and a_3 form an arithmetic sequence with common difference b_1, b_1, b_2 and b_3 form a geometric sequence with common ratio a_1, and $a_3 = b_3$. Find the minimum value of a_3, and determine the value of a_2b_2 when a_3 takes the minimum value.

Solution   Denote $a = a_1, b = b_1$, where $a, b > 0$.

Based on the question, we know $a_2 = a + b, a_3 = a + 2b, b_2 = ab, b_3 = a^2 b$. Denote $\lambda = a_3 = b_3$, and then $\lambda = a + 2b = a^2 b$. So $\lambda = a^2 b = (\lambda - 2b)^2 b$.

Using the AM-GM inequality, we get

$$4\lambda = (\lambda - 2b)^2 \cdot 4b \leqslant \left(\frac{2(\lambda - 2b) + 4b}{3} \right)^3 = \left(\frac{2\lambda}{3} \right)^3,$$

namely, $\lambda \geqslant \sqrt{\frac{27}{2}} = \frac{3\sqrt{6}}{2}$.

The equality holds when $\lambda - 2b = 4b$, i.e., $b = \frac{\lambda}{6}$. At this point λ (i.e., a_3) takes its minimum value $\frac{3\sqrt{6}}{2}$, and accordingly we have $b = \frac{\sqrt{6}}{4}, a = \lambda - 2b = \sqrt{6}$. Therefore,

$$a_2 b_2 = (a + b)ab = \frac{5\sqrt{6}}{4} \times \sqrt{6} \times \frac{\sqrt{6}}{4} = \frac{15\sqrt{6}}{8}.$$

11 **(20 marks)** Given that a and b are real numbers with $a < b$, the difference between the maximum value and the minimum value of function $y = \sin x$ on the closed interval $[a, b]$ is 1. Find the range of $b - a$.

Solution   Based on the graphical characteristics of the sine function, if $b - a > \pi$, then there exists a maximum or minimum point c and a zero point d of $\sin x$ within (a, b). Choose a sufficiently small positive number ε such that the interval $[d - \varepsilon, d + \varepsilon] \subseteq (a, b)$, where $\sin(d - \varepsilon)$ and $\sin(d + \varepsilon)$ have opposite signs. Thus, there exists $d_1 \in \{d + \varepsilon, d - \varepsilon\}$ such that $\sin d_1$ and $\sin c$ have opposite signs. Hence, $|\sin c - \sin d_1| > |\sin c| = 1$, which is a contradiction.

If $b - a < \frac{\pi}{3}$, then the maximum and minimum points of $\sin x$ on $[a, b]$ must be at the endpoints c and d of a monotonic interval $[c, d]$ of length less than $\frac{\pi}{3}$. However,

$$|\sin c - \sin d| = \left| 2 \sin \frac{c - d}{2} \cos \frac{c + d}{2} \right| \leqslant 2 \left| \sin \frac{c - d}{2} \right| < 2 \sin \frac{\pi}{6} = 1,$$

a contradiction.

On the other hand, let $a = 0, b = b_0 \in \left[\frac{\pi}{2}, \pi \right]$. Then the maximum value of $\sin x$ on $[a, b]$ is 1 and the minimum value is 0, which meets the requirements. At this time, $b - a$ can take any value in $\left[\frac{\pi}{2}, \pi \right]$.

Additionally, let $a = \arcsin(t - 1), b = \arcsin t$, where $t \in \left[\frac{1}{2}, 1 \right]$. Then the maximum value of $\sin x$ on $[a, b]$ is t and the minimum value is $t - 1$, which meets the requirements. When $t = \frac{1}{2}$, $b - a = \frac{\pi}{3}$; when $t = 1$, $b - a = \frac{\pi}{2}$. And $b - a$ changes continuously as t varies within $\left[\frac{1}{2}, 1 \right]$. Therefore, $b - a$ can take any value in $\left[\frac{\pi}{3}, \frac{\pi}{2} \right]$.

In summary, the range of $b - a$ is $\left[\frac{\pi}{3}, \frac{\pi}{2} \right] \cup \left[\frac{\pi}{2}, \pi \right] = \left[\frac{\pi}{3}, \pi \right]$.

2

China Mathematical Competition (Second Round)

2022

Test Paper A

1. (**40 marks**) As shown in Fig. 2.1, in convex quadrilateral $ABCD$, $\angle ABC = \angle ADC = 90°$. Point P on diagonal BD satisfies $\angle APB = 2\angle CPD$. Points X, Y on segment AP satisfy $\angle AXB = 2\angle ADB$, $\angle AYD = 2\angle ABD$. Prove that $BD = 2XY$.

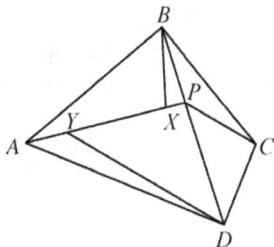

Fig. 2.1

Solution Note that $\angle ABC = \angle ADC = 90°$. As shown in Fig. 2.2, take the midpoint O of AC. Then O is the circumcenter of convex quadrilateral $ABCD$. It is obviously that P and B are on the same side of AC, otherwise $\angle APB \leqslant \angle CPD < 2\angle CPD$, which does not meet the problem statement.

13

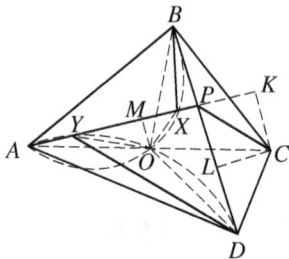

Fig. 2.2

According to the question, we have

$$\angle AXB = 2\angle ADB = \angle AOB,$$

$$\angle AYD = 2\angle ABD = \angle AOD.$$

By the above equations, we find that A, O, X, B are concyclic and A, Y, O, D are concyclic, respectively.

Therefore, $\angle OXA = \angle OBA = \angle CAB = \angle CDB, \angle OYP = \angle ODA = \angle CAD = \angle CBD$. Consequently, $\triangle OXY \backsim \triangle CDB$.

Let $OM \perp AP$ at point M, $CK \perp AP$ at point K, and $CL \perp BD$ at point L.

Since O is the midpoint of AC, we get $CK = 2OM$.

Since $\angle KPL = \angle APB = 2\angle CPD$, PC bisects $\angle KPL$, so $CK = CL$.

Considering that OM, CL are the altitudes on the corresponding sides XY, DB of similar triangles $\triangle OXY, \triangle CDB$, it follows that

$$\frac{XY}{BD} = \frac{OM}{CL} = \frac{OM}{CK} = \frac{1}{2},$$

and thus $BD = 2XY$.

2   **(40 marks)** Suppose integer $n(n > 1)$ has exactly k distinct prime factors. Denote the sum of all positive divisors of n by $\sigma(n)$. Prove that $\sigma(n)|(2n - k)!$

Solution   Let $n = \prod_{i=1}^{k} p_i^{\alpha_i}$ be the standard factorization of n.

Denote $m_i = 1 + p_i + \cdots + p_i^{\alpha_i} (i = 1, 2, \ldots, k)$, and then $\sigma(n) = \prod_{i=1}^{k} m_i$.

Proof 1 First, we prove that

$$2n - k \geqslant km_i(i = 1, 2, \ldots, k).$$ ①

In fact,

$$m = p_i^{\alpha_i} \left(1 + \frac{1}{p_i} + \cdots + \frac{1}{p_i^{\alpha_i}}\right) \leqslant p_i^{\alpha_i} \left(1 + \frac{1}{2} + \cdots + \frac{1}{2^{\alpha_i}}\right)$$

$$= p_i^{\alpha_i} \left(2 - \frac{1}{2^{\alpha_i}}\right) \leqslant 2p_i^{\alpha_i} - 1(i = 1, 2, \ldots, k).$$

Therefore,

$$m_i + 1 \leqslant 2p_i^{\alpha_i} = \frac{2n}{\prod_{\substack{j=1 \\ j \neq i}}^{k} p_j^{\alpha_j}} \leqslant \frac{2n}{2^{k-1}} \leqslant \frac{2n}{k},$$

where the last step holds since $2^{k-1} \geqslant 1 + \mathrm{C}_{k-1}^1 = k(k \geqslant 2)$ and $2^0 \geqslant 1$. Thus, inequality ① holds.

It can be seen from ① that for each $i = 1, 2, \ldots, k$, there are at least k multiples of m_i in $1, 2, \ldots, 2n - k$. Hence, mutually distinct positive integers $t_1, t_2, \ldots, t_k$ can be found in $1, 2, \ldots, 2n - k$, which are multiples of $m_1, m_2, \ldots, m_k$, respectively. Therefore, $\sigma(n) = \prod_{i=1}^{k} m_i$ divides $(2n - k)!$

Proof 2 Let $S_j = \sum_{i=1}^{j} m_i(j = 1, 2, \ldots, k)$, $S_0 = 0$. We prove the following two conclusions:

(1) $\sigma(n) \mid S_k!$;
(2) $S_k \leqslant 2n - k$.

Proof of Conclusion (1) For $i = 1, 2, \ldots, k$, among the m_i consecutive integers $S_{i-1} + 1, S_{i-1} + 2, \ldots, S_i$, there must be a multiple of m_i. Hence, $\frac{(S_{i-1}+1)(S_{i-1}+2)\cdots S_i}{m_i} \in \mathbf{Z}$.

Thus, $\prod_{i=1}^{k} \frac{(S_{i-1}+1)(S_{i-1}+2)\cdots S_i}{m_i} \in \mathbf{Z}$, which is equivalent to $\sigma(n) \mid S_k!$

Proof of Conclusion (2) For $i = 1, 2, \ldots, k$, there is

$$m_i = p_i^{\alpha_i} \left(1 + \frac{1}{p_i} + \cdots + \frac{1}{p_i^{\alpha_i}}\right)$$

$$\leqslant p_i^{\alpha_i} \left(1 + \frac{1}{2} + \cdots + \frac{1}{2^{\alpha_i}}\right)$$

$$= p_i^{\alpha_i} \left(2 - \frac{1}{2^{\alpha_i}}\right) \leqslant 2p_i^{\alpha_i} - 1.$$ ②

Denote $\lambda_i = p_i^{\alpha_i}\ (i = 1, 2, \ldots, k)$, and then $\lambda_i \geqslant 2$. Using 'if $a, b \geqslant 2$, then $ab \geqslant a + b$' repeatedly, we obtain

$$n = \prod_{i=1}^{k} \lambda_i \geqslant \sum_{i=1}^{k} \lambda_i.$$

Combining this with ②, we get

$$S_k = \sum_{i=1}^{k} m_i \leqslant \sum_{i=1}^{k} (2\lambda_i - 1)$$

$$= 2\sum_{i=1}^{k} \lambda_i - k \leqslant 2n - k.$$

By Conclusions (1) and (2), the original problem is proved.

③ (**50 marks**) Let $a_1, a_2, \ldots, a_{100}$ be non-negative integers such that

    (1) There exists positive integer $k \leqslant 100$ such that $a_1 \leqslant a_2 \leqslant \cdots \leqslant a_k$, and $a_i = 0$ for $i > k$;

    (2) $a_1 + a_2 + a_3 + \cdots + a_{100} = 100$;

    (3) $a_1 + 2a_2 + 3a_3 + \cdots + 100a_{100} = 2022$.

Find the minimum possible value of $a_1 + 2^2 a_2 + 3^2 a_3 + \cdots + 100^2 a_{100}$.

Solution 1   When $a_1 = a_2 = \cdots = a_{18} = 0, a_{19} = 19, a_{20} = 40, a_{21} = 41, a_{22} = a_{23} = \cdots = a_{100} = 0, k = 21$, the three given conditions are satisfied. At this time,

$$\sum_{i=1}^{100} i^2 a_i = 19^3 + 20^2 \times 40 + 21^2 \times 41 = 40940.$$

Next, we prove this is the minimum possible value.

First note $k \geqslant 21$. Otherwise, if $k \leqslant 20$, then

$$\sum_{i=1}^{100} i a_i = \sum_{i=1}^{k} i a_i \leqslant \sum_{i=1}^{k} 20 a_i \leqslant 2000,$$

which contradicts condition (3).

According to conditions (2) and (3), we have

$$\sum_{i=1}^{100} i^2 a_i = \sum_{i=1}^{100} (i-20)^2 a_i + 40 \sum_{i=1}^{100} i a_i - 400 \sum_{i=1}^{100} a_i$$

$$= \sum_{i=1}^{100} (i-20)^2 a_i + 40880.$$

When $a_{20} \leqslant 40$,

$$\sum_{i=1}^{100} (i-20)^2 a_i = \sum_{\substack{i=1, \\ i \neq 20}}^{100} (i-20)^2 a_i \geqslant \sum_{\substack{i=1, \\ i \neq 20}}^{100} a_i = 100 - a_{20} \geqslant 60.$$

Hence,

$$\sum_{i=1}^{100} i^2 a_i \geqslant 40940.$$

When $a_{20} \geqslant 41$, for $k \geqslant 21$ and condition (1) we have $a_{21} \geqslant 41$, and thus

$$\sum_{i=1}^{100} i^2 a_i = \sum_{i=1}^{100} (i-19)(i-20)a_i + 39 \sum_{i=1}^{100} i a_i - 380 \sum_{i=1}^{100} a_i$$

$$= \sum_{i=1}^{100} (i-19)(i-20)a_i + 40858$$

$$\geqslant (21-19)(21-20)a_{21} + 40858 \geqslant 40940.$$

In summary, the desired minimum value is 40940.

Solution 2 For non-negative integers $a_1, a_2, \ldots, a_{100}$ that satisfy the problem conditions, we can correspondingly take 100 positive integers $x_1, x_2, \ldots, x_{100} \in \{1, 2, \ldots, 100\}$, where there are exactly a_1 1's, a_2 2's, $\ldots, a_{100}$ 100's (condition (2) ensures that there are exactly 100 numbers). Conditions (1) and (3) are transformed into the following conditions (A) and (B), respectively:

(A) There exists a positive integer $k \leqslant 100$ such that $x_1, x_2, \ldots, x_{100}$ do not contain any number greater than k, and the number of 1's, the number of 2's, $\ldots$, the number of k's are increasing (not strictly);

(B) $\sum_{j=1}^{100} x_j = \sum_{i=1}^{100} i a_i = 2022$, i.e., the average of $x_1, x_2, \ldots, x_{100}$ is $\mu = 20.22$.

Noting that $\sum_{i=1}^{100} i^2 a_i = \sum_{j=1}^{100} x_j^2$, the problem is transformed into: 100 numbers

$$x_1, x_2, \ldots, x_{100} \in \{1, 2, \ldots, 100\}$$

satisfying conditions (A) and (B), find the minimum value of $\sum_{j=1}^{100} x_j^2$.

When $x_1, x_2, \ldots, x_{100}$ takes 19 19's, 40 20's, and 41 21's,

$$\sum_{j=1}^{100} x_j^2 = 40940.$$

Now let us prove that $\sum_{j=1}^{100} x_j^2$ is at least 40940.
Since

$$\sum_{j=1}^{100} x_j^2 = \sum_{j=1}^{100} (x_j - \mu)^2 - 100\mu^2 + 2\mu \sum_{j=1}^{100} x_j = 100\mu^2 + \sum_{j=1}^{100} (x_j - \mu)^2,$$

it is converted into considering the minimum value of $\sum_{j=1}^{100} (x_j - \mu)^2$.

From $\mu = 20.22$, we know that there exist $x_j \geqslant 21$ and $x_j \leqslant 20$. Suppose there are a of $x_j \geqslant 21$, b of $x_j = 20$, and c of $x_j \leqslant 19$ in $x_1, x_2, \ldots, x_{100}$. According to condition (A), $a \geqslant b$.

We relax condition (A) to (A'): $a \geqslant b$. Under conditions (A') and (B), we will prove that the minimum value is still obtained when there are 19 19's, 40 20's, and 41 21's.

Since there are limited ways to select $x_1, x_2, \ldots, x_{100}$ that satisfy (A') and (B), we select the group of $x_1, x_2, \ldots, x_{100}$ that minimizes the sum of squares.

If $c \geqslant 19$, noting that $a + b + c = 100$ and $a \geqslant b$, there are

$$\sum_{j=1}^{100} (x_j - \mu)^2 \geqslant 0.78^2 a + 0.22^2 b + 1.22^2 c$$

$$\geqslant 0.78^2 \cdot \left\lceil \frac{100 - c}{2} \right\rceil + 0.22^2 \cdot \left\lfloor \frac{100 - c}{2} \right\rfloor + 1.22^2 c$$

$$\geqslant 0.78^2 \times 41 + 0.22^2 \times 40 + 1.22^2 \times 19.$$

If $c \leqslant 18$, then $a + b \geqslant 82$. At this time, there is $c > 0$, because if $c = 0$, then the average of x_j is not less than 20.5, which is inconsistent with condition (B).

We also have $b > 0$. Otherwise, if $b = 0$, then from $a \geqslant 82$ and $c > 0$, we can take one $x_i < 20$ and one $x_j > 20$. Replacing them with $x_i + 1$ and $x_j - 1$, the average remains unchanged but

$$(x_i + 1)^2 + (x_j - 1)^2 < x_i^2 + x_j^2,$$

which indicates that the sum of squares becomes smaller. With a decreasing by at most 1, and b increasing by at most 2, conditions (A') and (B) are still satisfied, which contradicts the fact that $x_1, x_2, \ldots, x_{100}$ minimize the sum of squares.

And if there is one $x_i \leqslant 18$, then $b > 0$ can be used to choose one $x_j = 20$. Replacing x_i, x_j with $x_i + 1$ and $x_j - 1$, the average remains unchanged but the sum of squares decreases. With b decreasing by 1, conditions (A') and (B) are still satisfied, which contradicts the fact that $x_1, x_2, \ldots, x_{100}$ minimize the sum of squares.

So all c's $x_j \leqslant 19$ are equal to 19. But at this point

$$\sum_{j=1}^{100} (x_j - \mu) \geqslant 0.78a - 0.22b - 1.22c$$

$$\geqslant 0.78 \cdot \left\lceil \frac{100 - c}{2} \right\rceil - 0.22 \cdot \left\lceil \frac{100 - c}{2} \right\rceil - 1.22c$$

$$\geqslant 0.78 \times 41 - 0.22 \times 41 - 1.22 \times 18 > 0,$$

contradicting condition (B).

So if and only if $x_1, x_2, \ldots, x_{100}$ are 19 19's, 40 20's, and 41 21's, $\sum_{j=1}^{100} (x_j - \mu)^2$ takes the minimum value. Accordingly, $\sum_{i=1}^{100} i^2 a_i = \sum_{j=1}^{100} x_j^2$ takes the minimum value 40940.

4 **(50 marks)** Find the smallest positive integer t that satisfies the following properties: In a 100×100 grid, if each small square is colored with a certain color and the number of small squares of each color does not exceed 104, then there exists a $1 \times t$ or $t \times 1$ rectangle such that t contain small squares of at least three different colors.

Solution Divide the grid into 100 squares of 10×10. Color the 100 small squares in each square with the same color, and different squares are colored with different colors. This coloring method satisfies the problem's condition, and it is easy to see that any 1×11 or 11×1 rectangle contains small squares with at most two different colors. Therefore, $t \geqslant 12$.

Next, we will prove that $t = 12$ has the described property. We need the following lemma.

Lemma   *In a 1×100 grid X, if each small square is colored with a certain color and if one of the following two conditions is met, then there exists a 1×12 rectangle such that it contains at least three colors.*

(1) *There are at least 11 colors in X.*

(2) *There are exactly 10 colors in, and each color appears in exactly 10 small squares.*

Proof of the lemma   We use proof by contradiction. Assume the conclusion does not hold.

Take the rightmost small square of each color. Let these squares be at (from left to right) $x_1 < x_2 < \cdots < x_k$, with colors $c_1, c_2, \ldots, c_k$, respectively. Then for $2 \leqslant i < k$, we have $x_i - x_{i-1} \geqslant 11$. This is because if $x_i - x_{i-1} \leqslant 10$, then squares from x_{i-1} to $x_i + 1$ (not exceeding 12 squares) will contain at least three different colors (square x_{i-1} is color c_{i-1}, square x_i is color c_i, and the color of square $x_i + 1$ is different from c_{i-1} and c_i), which contradicts the assumption.

If condition (1) holds, then $k \geqslant 11$, and thus $x_{10} \geqslant x_1 + 9 \times 11 \geqslant 100, x_{11} > 100$, which is a contradiction. Therefore, the conclusion holds under condition (1).

If condition (2) holds, consider squares from x_{i-1} to x_{i-1}. Because each color appears in at most 10 squares, the 11 squares must contain at least two different colors, both of which are different from c_1. Thus, squares from x_1 to $x_1 + 11$ contains at least three colors, contradicting condition (2). Therefore, the conclusion holds under condition (2) as well.

The lemma is proved.

Returning to the original problem, let $c_1, c_2, \ldots, c_k$ be all colors that appear.

For $1 \leqslant i \leqslant k$, let s_i be the number of small squares of color c_i, u_i be the number of rows containing color c_i, and v_i be the number of columns containing color c_i. From the conditions, we know $s_i \leqslant 104$. It is also clear that $u_i v_i \geqslant s_i$, and the equality sign holds if and only if all small squares of color c_i are at the intersections of the rows and columns containing color c_i.

In the following we will prove $u_i + v_i \geqslant \frac{s_i}{5}$, where the equality holds if and only if $u_i = v_i = 10, s_i = 100$.

If $u_i + v_i \geqslant 21$, then from $s_i \leqslant 104$ we know that $u_i + v_i > \frac{1}{5} s_i$; if $u_i + v_i \leqslant 20$, then

$$u_i + v_i \geqslant \frac{(u_i + v_i)^2}{20} \geqslant \frac{u_i v_i}{5} \geqslant \frac{s_i}{5},$$

the equality holds if and only if $u_i = v_i = 10, s_i = 100$.

Therefore, $\sum_{i=1}^{k} (u_i + v_i) \geqslant \sum_{i=1}^{k} \frac{1}{5} s_i$.

If $\sum_{i=1}^{k} (u_i + v_i) > 2000$, from the pigeonhole principle, there exists a row or column containing at least 11 colors.

If $\sum_{i=1}^{k} (u_i + v_i) = 2000$, then by the condition that the equality holds, it follows that each color must color 100 squares and these squares are located at the intersections of 10 rows and 10 columns. Therefore, there are exactly 10 colored squares in each row and column, and there are exactly 100 squares of each color.

From the lemma, we know that both situations lead to the existence of a 1×12 or 12×1 rectangle containing small squares of at least three colors.

In summary, the desired smallest t is 12.

Test Paper B

1 **(40 marks)** As shown in Fig. 2.3, suppose four points A, B, C, D are arranged in order on circle ω, with AC passing through its center O. Point P on segment BD satisfies $\angle APC = \angle BPC$. Points X, Y on segment AP satisfy that points A, O, X, B are concyclic, and points A, Y, O, D are concyclic. Prove that $BD = 2XY$.

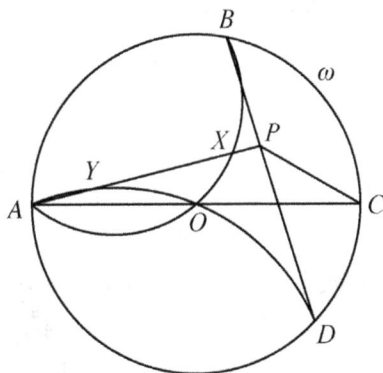

Fig. 2.3

Solution Based on the problem statement, we have

$$\angle OXA = \angle OBA = \angle CAB = \angle CDB,$$

$$\angle OYP = \angle ODA = \angle CAD = \angle CBD,$$

and thus $\triangle OXY \backsim \triangle CDB$.

As shown in Fig. 2.4, let $OM \perp AP$ at point M, and $CK \perp AP$ at point K (with K on circle ω), and $CL \perp BD$ at point L.

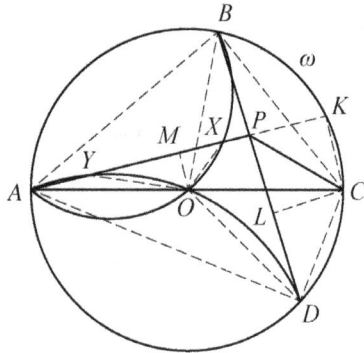

Fig. 2.4

Since O is the midpoint of AC, we have $CK = 2OM$. From $\angle APC = \angle BPC$, we get $CK = CL$.

Considering that OM, CL are the altitudes on the corresponding sides XY, BD of similar triangles $\triangle OXY, \triangle CDB$, respectively, it follows that

$$\frac{XY}{BD} = \frac{OM}{CL} = \frac{OM}{CK} = \frac{1}{2},$$

and thus $BD = 2XY$.

2 **(40 marks)** Given positive real numbers a and b with $a < b$. Let $x_1, x_2, \ldots, x_{2022} \in [a, b]$. Find the maximum value of

$$\frac{|x_1 - x_2| + |x_2 - x_3| + \cdots + |x_{2021} - x_{2022}| + |x_{2022} - x_1|}{x_1 + x_2 + \cdots + x_{2022}}.$$

Solution First, we prove that for $x, y \in [a, b]$, there is

$$|x - y| \leqslant \frac{b - a}{a + b}(x + y). \qquad \text{①}$$

Without loss of generality, assume $a \leqslant x \leqslant y \leqslant b$. Then $\frac{a}{b} \leqslant \frac{x}{y} \leqslant 1$, and thus

$$\frac{|x - y|}{x + y} = \frac{y - x}{y + x} = \frac{1 - \frac{x}{y}}{1 + \frac{x}{y}} \leqslant \frac{1 - \frac{a}{b}}{1 + \frac{a}{b}} = \frac{b - a}{b + a}.$$

Therefore, inequality ① is proved.

Consequently,

$$|x_1 - x_2| + |x_2 - x_3| + \cdots + |x_{2021} - x_{2022}| + |x_{2022} - x_1|$$

$$\leqslant \frac{b-a}{a+b}((x_1 + x_2) + (x_2 + x_3) + \cdots + (x_{2022} + x_1))$$

$$= \frac{2(b-a)}{a+b}(x_1 + x_2 + \cdots + x_{2022}),$$

Hence,

$$\frac{|x_1 - x_2| + |x_2 - x_3| + \cdots + |x_{2021} - x_{2022}| + |x_{2022} - x_1|}{x_1 + x_2 + \cdots + x_{2022}} \leqslant \frac{2(b-a)}{a+b},$$

where the equality holds when $x_1 = x_3 = \cdots = x_{2021} = a, x_2 = x_4 = \cdots = x_{2022} = b$.

Therefore, the desired maximum value is $\frac{2(b-a)}{a+b}$.

3 **(50 marks)** Let positive integers a, b each have exactly $m(m \geqslant 3)$ positive divisors, with $a_1, a_2, a_3, \ldots, a_m$ being a permutation of all the positive divisors of a. Is it possible that $a_1, a_1 + a_2, a_2 + a_3, \ldots, a_{m-1} + a_m$ could be exactly a permutation of all the positive divisors of b? Prove your conclusion.

Solution The answer is negative.

We will use proof by contradiction. Assume that the situation described in the question occurs, and denote

$$A = \{a_1, a_2, a_3, \ldots, a_m\}, \quad B = \{a_1, a_1 + a_2, a_2 + a_3, \ldots, a_{m-1} + a_m\}.$$

It is obvious that $a_i + a_{i+1} \geqslant 3(i = 1, 2, \ldots, m - 1)$. However, since $1 \in B$, we have $a_1 = 1$. Also, $2 \notin B$, so b is odd.

Therefore, $a_1 + a_2$ is odd, implying a_2 is even. Since $a_2 \in A$, a is even. It is easy to know that the two largest elements in A are a and $\frac{a}{2}$.

Obviously, every element in B does not exceed $a + \frac{a}{2} = \frac{3a}{2}$. In particular, there is $b \leqslant \frac{3}{2}a$.

Suppose $a_i = a, a_j = \frac{a}{2}$, where $i, j \geqslant 2$ (since a has $m(m \geqslant 3)$ positive divisors, and $a_1 = 1$). Thus, B contains two elements $a_{i-1} + a_i, a_{j-1} + a_j$, both greater than $\frac{a}{2}$ and hence both greater than $\frac{b}{3}$, and both are factors of b. This shows that $2 \mid b$, contradicting the fact that b is odd.

Therefore, the given situation cannot occur.

4 **(50 marks)** There are 100 points $P_1, P_2, \ldots, P_{100}$ in order on the circumference, where P_{100} and P_1 are adjacent. Now there are 5 colors, and each point in $P_1, P_2, \ldots, P_{100}$ is required to be colored in one of the 5 colors, with each color being used at least once. For any such coloring, it is required that among $P_1, P_2, \ldots, P_{100}$, there must be at least t consecutive points containing at least 3 different colors. Find the minimum value of t.

Solution First, let $P_{25}, P_{50}, P_{75}, P_{100}$ be colored in colors $1, 2, 3, 4$, respectively, and the remaining points in color 5. In this case, any 25 consecutive points among $P_1, P_2, \ldots, P_{100}$ contain only two colors, which does not meet the requirement. Therefore, t must be at least 26.

Method 1 In the following, assume that there is a coloring method such that any 26 consecutive points contain at most two colors. For this coloring, choose as many consecutive points as possible from P such that these points do not contain all 5 colors, thus containing exactly 4 colors (Otherwise, we can add an adjacent point of these consecutive points, still not containing all 5 colors). Assume the selected points are $P_{k+1}, P_{k+2}, \ldots, P_{100}$, and these $100 - k$ points do not contain color 1. By extremity, it is known that P_1 and P_k are both colored in color 1.

For $i = 1, 2, 3, 4, 5$, let x_i denote the maximum index of the points colored in color i, and then $x_1 = k$.

By symmetry, we can assume $x_2 < x_3 < x_4 < x_5$. Thus $x_5 = 100$, and it is easy to see $x_2 > k$.

For each $i = 2, 3, 4, 5$, we prove $x_i - x_{i-1} \geqslant 25$.

In fact, if $x_i - x_{i-1} \leqslant 24$, consider the 26 consecutive points $P_{x_{i-1}}, P_{x_{i-1}+1}, \ldots, P_{x_{i-1}+25}$ (indices taken modulo 100), where $P_{x_{i-1}}$ is colored in $i - 1$, P_{x_i} is colored in i, and P_{x_i+1} is colored in color j different from $i - 1$ and i (when $i = 2, 3, 4$, there must be $i + 1 \leqslant j \leqslant 5$; when $i = 5$, since $P_{x_5+1} = P_{101} = P_1$, there must be $j = 1 \notin \{4, 5\}$), which means these points contain at least 3 colors, contradicting the assumption.

Therefore, $x_5 = x_1 + \sum_{i=2}^{5} (x_i - x_{i-1}) \geqslant 1 + 4 \times 25 > 100$, which contradicts $x_5 = 100$.

The above shows that for any valid coloring, there are always 26 consecutive points among $P_1, P_2, \ldots, P_{100}$ containing at least 3 colors.

To sum up, the desired minimum value of t is 26.

Method 2 Consider any coloring method. For $i = 1, 2, 3, 4, 5$, let n_i denote the number of points colored in color 1 among $P_1, P_2, \ldots, P_{100}$, where $n_1 + n_2 + \cdots + n_5 = 100$.

Consider the number of arcs $\gamma_k = \overset{\frown}{P_k P_{k+1} \cdots P_{k+25}}$ $(k = 1, 2, \ldots, 100)$ containing color 1, denoted as s_1, where the subscripts are understood modulo 100.

If any 26 consecutive points contain color 1, then $s_1 = 100$.

If there exist 26 consecutive points that do not contain color 1, we might assume that all n_1 points colored in color 1 are $P_{i_1}, P_{i_2}, \ldots, P_{i_{n_1}}$ $(27 \leqslant i_1 < i_2 < \cdots < i_{n_1} \leqslant 100)$, and then

$$\gamma_{i_1-25}, \gamma_{i_1-24}, \ldots, \gamma_{i_1-1}, \gamma_{i_1}, \gamma_{i_2}, \ldots, \gamma_{i_{n_1}}$$

are $n_1 + 25$ distinct arcs. Thus, $s_1 \geqslant n_1 + 25$.

This implies $s_1 \geqslant \min\{n_1 + 25, 100\}$.

Similarly, for $i = 2, 3, 4, 5$, the number s_i of arc $\gamma_k (k = 1, 2, \ldots, 100)$ containing color i satisfies $s_i \geqslant \min\{n_i + 25, 100\}$.

Therefore, $s_1 + s_2 + \cdots + s_5 \geqslant S$, where we denote

$$S = \sum_{i=1}^{5} \min\{n_i + 25, 100\}.$$

If there exists a number in $n_i (1 \leqslant i \leqslant 5)$ that is not less than 75, we might as well set $n_1 \geqslant 75$, and then

$$S \geqslant 100 + \sum_{i=2}^{5} (1 + 25) > 200.$$

If all $n_i (1 \leqslant i \leqslant 5)$ are less than 75, then

$$s_1 + s_2 + \cdots + s_5 \geqslant \sum_{i=1}^{5} (n_i + 25) = 100 + 5 \times 25 > 200.$$

Therefore,

$$s_1 + s_2 + \cdots + s_5 \geqslant S \geqslant 201.$$

By the pigeonhole principle, there must be at least one arc corresponding to at least

$$\left\lceil \frac{201}{100} \right\rceil = 3$$

different colors, and the 26 points covered by this arc contain at least 3 different colors.

To sum up, the desired minimum value of t is 26.

3

China Mathematical Olympiad

The 2022 China Mathematical Olympiad (also named the 38th National Mathematics Winter Camp for Middle School Students) was held on 29th and 30th December 2022. This examination was organized by the Science and Technology Associations and Mathematics Societies of each province, autonomous region, and municipality in the provincial capitals across the country. (In many provinces, students with fever symptoms were arranged to take exams separately.) Online invigilation was organized by the hosting organizer, Shenzhen Middle School. The event is the highest level and most qualified mathematics competition for high school students in China.

This winter camp included representative teams from 31 provinces, autonomous regions, and municipalities, Hong Kong Special Administrative Region of China, Macau Special Administrative Region of China, as well as an international team from Singapore. There are a total of 608 participants, of which 60 students qualified for the China National Training Team.

First Day
(8:00–12:30; 29th December 2022)

1 Suppose positive real number sequences a_n, b_n satisfy: for any positive integer n, there are

$$a_{n+1} = a_n - \frac{1}{1 + \sum_{i=1}^{n} \frac{1}{a_i}}, \quad b_{n+1} = b_n + \frac{1}{1 + \sum_{i=1}^{n} \frac{1}{b_i}}.$$

27

(1) If $a_{100}b_{100} = a_{101}b_{101}$, find the value of $a_1 - b_1$.

(2) If $a_{100} = b_{99}$, which one of $a_{100} + b_{100}$ and $a_{101} + b_{101}$ is larger?

(Contributed by He Yijie)

Solution (1) Let $a_1 = a$, $b_1 = b$ $(a, b > 0)$. According to the problem, for any positive integer $n \geqslant 2$, there is

$$\frac{1}{a_n - a_{n+1}} = 1 + \sum_{i=1}^{n} \frac{1}{a_i},$$

and thus

$$\frac{1}{a_n - a_{n+1}} = \frac{1}{a_{n-1} - a_n} + \frac{1}{a_n}.$$

That is, $\frac{a_n}{a_n - a_{n+1}} = \frac{a_n}{a_{n-1} - a_n} + 1 = \frac{a_{n-1}}{a_{n-1} - a_n}$. This leads to $\frac{a_{n+1}}{a_n} = \frac{a_n}{a_{n-1}}$.

From the given conditions, we have $a_2 = a_1 - \frac{1}{1 + \frac{1}{a_1}} = \frac{a}{a+1}a_1$. Thus,

$$\frac{a_{n+1}}{a_n} = \frac{a_n}{a_{n-1}} = \cdots = \frac{a_2}{a_1} = \frac{a}{a+1}.$$

Hence, the common ratio of sequence $\{a_n\}_{n=1}^{\infty}$ is $\frac{a}{a+1}$, and the general term is $a_n = \frac{a^n}{(a+1)^{n-1}}$.

Similarly, transforming the general term of b_n: for any positive integer $n \geqslant 2$, there is

$$\frac{1}{b_{n+1} - b_n} = 1 + \sum_{i=1}^{n} \frac{1}{b_i} = \frac{1}{b_n - b_{n-1}} + \frac{1}{b_n}.$$

That is,

$$\frac{b_n}{b_{n+1} - b_n} = \frac{b_n}{b_n - b_{n-1}} + 1 = \frac{b_{n-1}}{b_n - b_{n-1}} + 2$$

$$= \cdots = \frac{b_1}{b_2 - b_1} + 2(n-1) = b + 2n - 1.$$

Hence, $\frac{b_{n+1}}{b_n} = \frac{b+2n}{b+2n-1}$.

Therefore, the general term of $\{b_n\}_{n=1}^{\infty}$ is

$$b_n = b \prod_{k=1}^{n-1} \frac{b+2k}{b+2k-1}.$$

If $a_{100}b_{100} = a_{101}b_{101}$, namely $\frac{a_{100}}{a_{101}} = \frac{b_{101}}{b_{100}}$, then $\frac{a+1}{a} = \frac{b+200}{b+199}$. Solving this gives $a = b + 199$. Therefore, $a_1 - b_1 = a - b = 199$.

(2) Method 1: By definition, $\{a_n\}_{n=1}^{\infty}$ is monotonically decreasing, and $\{b_n\}_{n=1}^{\infty}$ is monotonically increasing. From the condition $a_{100} = b_{99}$, we know

$$a_1 > a_2 > \cdots > a_{99} > a_{100} = b_{99} > b_{98} > \cdots > b_1 > 0. \qquad \text{①}$$

And then it follows that

$$a_{99} = a_{100} + \frac{1}{1 + \sum_{i=1}^{99} \frac{1}{a_i}} > b_{99} + \frac{1}{1 + \sum_{i=1}^{99} \frac{1}{b_i}} = b_{100}.$$

Combining this with ①, it follows that

$$a_{100} - a_{101} = \frac{1}{1 + \frac{1}{a_1} + \cdots + \frac{1}{a_{98}} + \frac{1}{a_{99}} + \frac{1}{a_{100}}}$$

$$> \frac{1}{1 + \frac{1}{b_1} + \cdots + \frac{1}{b_{98}} + \frac{1}{b_{100}} + \frac{1}{b_{99}}} = b_{101} - b_{100}.$$

Therefore, we get $a_{100} + b_{100} > a_{101} + b_{101}$.

Method 2: By $a_{100} = b_{99}$, it follows that

$$\frac{a^{100}}{(a+1)^{99}} = \frac{b(b+2)\cdots(b+196)}{(b+1)(b+3)\cdots(b+195)}. \qquad \text{②}$$

We prove $a_{100} + b_{100} > a_{101} + b_{101}$, which is equivalent to

$$a_{100} - a_{101} > b_{101} - b_{100}$$

$$\Leftrightarrow \frac{a_{100} - a_{101}}{a_{100}} > \frac{b_{101} - b_{100}}{b_{99}}$$

$$\Leftrightarrow 1 - \frac{a}{a+1} > \frac{(b+198)(b+200)}{(b+197)(b+199)} - \frac{b+198}{b+197}$$

$$\Leftrightarrow \frac{1}{a+1} > \frac{b+198}{(b+197)(b+199)}$$

$$\Leftrightarrow a < b + 197 - \frac{1}{b+198}. \qquad \text{③}$$

We use proof by contradiction. Assume $a \geqslant b + 197 - \frac{1}{b+198}$. Then it is clear that $a > b + 196$, and thus

$$\frac{a^2}{a+1} - (b+196)$$

$$= \frac{a(a-b-196) - (b+196)}{a+1}$$

$$\geqslant \frac{1}{a+1}\left(\left(b+197-\frac{1}{b+198}\right)\left(1-\frac{1}{b+198}\right)-(b+196)\right)$$

$$> \frac{1}{a+1}\left(\left(b+197-\frac{1}{b+197}\right)\frac{b+197}{b+198}-(b+196)\right)$$

$$= 0.$$

And since $\frac{a}{a+1} > \frac{b+194}{b+195} > \frac{b+192}{b+193} > \cdots > \frac{b}{b+1}$, we get

$$\frac{a^{100}}{(a+1)^{99}} = \left(\frac{a}{a+1}\right)^{98}\frac{a^2}{a+1} > \frac{b}{b+1}\cdots\frac{b+194}{b+195}(b+196),$$

which contradicts ②. Therefore, ③ holds, and thus $a_{100}+b_{100} > a_{101}+b_{101}$.

Method 3: In the following, we give another proof of $a < b+197-\frac{1}{b+198}$. By $a_{100} = b_{99}$ and $a_n - a_{n+1} < 1$, $b_{n+1} - b_n < 1(1 \leqslant n \leqslant 99)$, it follows that

$$a - b = \sum_{i=1}^{98}(b_{i+1}-b_i) + \sum_{i=1}^{99}(a_i-a_{i+1})$$

$$\leq 196 + (b_2-b_1) = 196 + \frac{b_1}{b_1+1}$$

$$= 197 - \frac{1}{b+1} < 197 - \frac{1}{b+198}.$$

Method 4: In the following, we give another proof of $a < b+197-\frac{1}{b+198}$. On one hand, by Bernoulli's inequality, we have

$$a - 99 < a\left(1-\frac{99}{a+1}\right) < a\left(1-\frac{1}{a+1}\right)^{99}$$

$$= a\left(\frac{a}{a+1}\right)^{99} = \frac{a^{100}}{(a+1)^{99}}.$$

On the other hand,

$$\frac{b(b+2)\cdots(b+196)}{(b+1)(b+3)\cdots(b+195)}$$

$$= b\left(1+\frac{1}{b+1}\right)\left(1+\frac{1}{b+3}\right)\cdots\left(1+\frac{1}{b+195}\right)$$

$$\leqslant b\left(1 + \frac{1}{b+1}\right)\left(1 + \frac{1}{b+2}\right)\cdots\left(1 + \frac{1}{b+98}\right)$$

$$= \frac{b(b+99)}{b+1} = b + 98 - \frac{98}{b+1}.$$

Combined with ②, we know $a < b + 197 - \frac{98}{b+1} < b + 197 - \frac{1}{b+198}$.

Remark This problem examines the basic skills in algebra. Question (1) tests the ability to find the general term of a sequence, while question (2) assesses the fundamental technique of inequality approximation. Both questions are relatively independent of each other. Some students failed to find the general term, while they solved question (2).

Finding the general term of $\{b_n\}_{n=1}^\infty$ in question (1) was slightly more difficult. Some students had a less than ideal grasp of mathematical induction. After guessing the general term of $\{b_n\}_{n=1}^\infty$, if the basic steps and explanations of induction are missing, 3 points will be deducted. Question (2) contains a pitfall: when $a \geqslant b + 196$, many students incorrectly used the inequality $\frac{a}{a+1} > \frac{b+2i}{b+2i+1}$, $0 \leqslant i \leqslant 97$, after multiplying the 98 terms and then mistakenly asserting (the number of terms on the left and right sides of the inequality does not match)

$$\frac{a^{100}}{(a+1)^{99}} > \frac{b}{b+1}\cdots\frac{b+194}{b+195}(b+196).$$

In fact, this inequality does not hold when $a = b + 196$. After general decomposition, comparing the coefficients of the second highest term of b shows the contradiction. We hope students will pay closer attention to these concepts in future studies and practice.

The scores for this problem are as follows:

Score	21	18	15	12	9	6	3	0
Count	364	22	24	148	37	6	6	1

The average score for this problem is 17.4 points. Among students who received silver medals or higher, the average score is 18.8 points. (However, among the 60 students who entered the National Training Team, the average score did not reach 20.)

2 Given equilateral triangle ABC with side length 1, ($\triangle DEF$, $\triangle XYZ$) is called a *good triangle pair* if points D, E, F lie in the interior of segments BC, CA, and AB, respectively, and points X,

Y, Z lie on lines BC, CA, and AB, respectively, satisfying

$$\frac{DE}{20} = \frac{EF}{22} = \frac{FD}{38}, \quad \text{and} \quad DE \perp XY, \ EF \perp YZ, \ FD \perp ZX$$

(see Fig. 3.1). As $(\triangle DEF, \triangle XYZ)$ run through all good triangle pairs, determine all possible values of $\frac{1}{S_{\triangle DEF}} + \frac{1}{S_{\triangle XYZ}}$.

(Contributed by Yao Yijun)

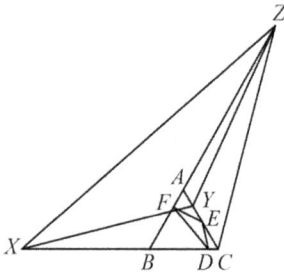 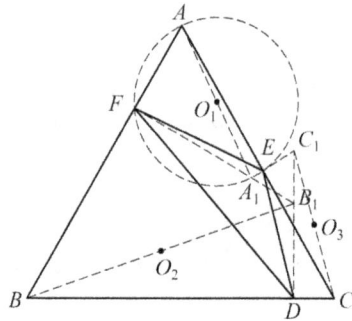

Fig. 3.1   Fig. 3.2

Solution

(1) First, consider the rotation of $90°$ (clockwise or anticlockwise), centered at an arbitrary point on the plane. Then, the images of points X, Y, Z, denoted as X', Y', Z', satisfy

$$X'Y' \parallel DE, \ Y'Z' \parallel EF, \ Z'X' \parallel FD.$$

In this case, $\triangle XYZ$ and $\triangle DEF$ are directly similar. Thus, *every good triangle pair is directly similar.*

(2) As shown in Fig. 3.2, rotate $\triangle XYZ$ (together with equilateral triangle ABC) by $90°$ and properly rescale and translate the picture so that the image of $\triangle XYZ$ coincides with $\triangle DEF$. At this point, points A, B, C are transformed into points A_1, B_1, C_1. Therefore, there are three points A_1, B_1, C_1 on the plane, satisfying the following:

- $\triangle A_1 B_1 C_1$ is an equilateral triangle.
- Points D, E, F lie on lines $B_1 C_1$, $C_1 A_1$, $A_1 B_1$, respectively.
- $A_1 B_1 \perp AB$, $B_1 C_1 \perp BC$, $C_1 A_1 \perp CA$.

From this, we see that points A_1, B_1, C_1 lie on the circumcircles of $\triangle AEF$, $\triangle BFD$, $\triangle CDE$, respectively. Moreover, A_1, B_1, C_1 are the *antipodal points* of A, B, C in the corresponding figure.

(3) Note that $S_{\triangle ABC} : S_{\triangle XYZ} = S_{\triangle A_1 B_1 C_1} : S_{\triangle DEF}$. So we only need to calculate the ratio of $S_{\triangle ABC} + S_{\triangle A_1 B_1 C_1}$ to $S_{\triangle DEF}$. We have

$$S_{\triangle ABC} = S_{\triangle DEF} + S_{\triangle EAF} + S_{\triangle FBD} + S_{\triangle DCE},$$

$$S_{\triangle A_1 B_1 C_1} = S_{\triangle DEF} + S_{\triangle EA_1 F} + S_{\triangle FB_1 D} + S_{\triangle DC_1 E}.$$

Here the right-hand sides are expressed in terms of oriented areas. *Since the two equilateral triangles on the left-hand side are directly similar, the signs on the areas are the same.* Using the properties of the above-mentioned antipodal points, if we denote the circumcentres of $\triangle AEF$, $\triangle BFD$, $\triangle CDE$ as O_1, O_2, O_3, respectively, then the condition that 'D, E, F lie on the interior of three sides' guarantees that these three circumcentres lie outside of $\triangle DEF$. Therefore, we have the following equality of areas:

$$S_{\triangle ABC} + S_{\triangle A_1 B_1 C_1} = 2(S_{\triangle DEF} + S_{\triangle EO_1 F} + S_{\triangle FO_2 D} + S_{\triangle DO_3 E}).$$

In fact, $\triangle O_1 O_2 O_3$ is the outer Napoleon triangle of $\triangle DEF$, and its area is half of the quantity in parentheses on the right side of the equality.

(4) So we need to calculate, for a triangle with side length ratio $20 : 22 : 38$, the ratio of the area of the hexagon formed by the vertices of the triangle and the vertices of the outer Napoleon triangle, to the area of the original triangle.

By Heron's formula, the area of a triangle with side lengths $20, 22, 38$ is

$$\sqrt{40 \cdot (40 - 20) \cdot (40 - 22) \cdot (40 - 38)} = 120\sqrt{2}.$$

On the other hand,

$$S_{\triangle EO_1 F} + S_{\triangle FO_2 D} + S_{\triangle DO_3 E} = \frac{\sqrt{3}}{3} \cdot (11^2 + 10^2 + 19^2) = 194\sqrt{3}.$$

Therefore,

$$\frac{1}{S_{\triangle DEF}} + \frac{1}{S_{\triangle XYZ}} = \frac{1}{S_{\triangle ABC}} \left(\frac{S_{\triangle ABC}}{S_{\triangle DEF}} + \frac{S_{\triangle ABC}}{S_{\triangle XYZ}} \right)$$

$$= \frac{1}{S_{\triangle ABC}} \times \frac{S_{\triangle ABC} + S_{\triangle A_1 B_1 C_1}}{S_{\triangle DEF}}$$

$$= \frac{4}{\sqrt{3}} \times \frac{2(120\sqrt{2} + 194\sqrt{3})}{120\sqrt{2}}$$

$$= \frac{97\sqrt{2} + 40\sqrt{3}}{15}.$$

Remark First, the good triangle pair described in this problem does exist. It is easy to find three points on the lines where the three sides of a given triangle (in this case, $\triangle ABC$) lie, such that the resulting triangle has the given shape and the direction of one of the sides is specified (and thus the direction of all three sides is determined).

At the end of the question, there are the reciprocals of the areas of the triangles. We know that in classical plane geometry, the only meaningful quantity is the ratio of length to area, so we have to think of comparing the area of the two triangles to a 'base' triangle (in this case, $\triangle ABC$).

The geometric transformations involved in this problem are not complex, but the directed area part requires some conceptual understanding. The key part of the calculation: for given $\triangle DEF$, the sum of the areas of two equilateral triangles $\triangle ABD$ and $\triangle A_1 B_1 C_1$ with corresponding sides perpendicular to each other is a constant. This theorem is attributed to the Italian mathematician Ernest Cesàro.[1]

The vast majority of students who solved this problem used a more computationally intensive method:

- On the basis of the similarity of two triangles in the same orientation, find their rotational similarity center.
- Prove that for good triangle pairs, this similarity center is the same point.
- The right triangle formed by the lines connecting corresponding points on these two triangles to this center is similar.
- Then, using an equivalent form of the Pythagorean theorem, we have the following: 'The sum of the reciprocals of the squares of the two vertical sides of a right-angled triangle is equal to the reciprocal of the square of the altitude on the hypotenuse,' the reciprocals of the areas of these two triangles transform into the reciprocal of the area of a triangle.
- Finally, the problem reduces to the following: find a point such that its perpendicular distances to the three sides of an equilateral triangle are in the ratio 20 : 22 : 38, and then find the reciprocal of the area of this triangle.

There were also a small number of candidates who took the approach of expressing the area of a triangle directly as a trigonometric function of the slope of one of the sides of $\triangle DEF$ (or $\triangle XYZ$) (i.e., the angle between one of their sides and one of the sides of the equilateral triangle). They found

---

[1]This theorem was published in 1883 in the journal *Mathesis*, edited by P. Mansion and J. Neuberg. At that time, Ernest Cesàro (1859–1906) was studying at the Mining Academy in Liège, Belgium (where his mathematics professor was Catalan).

the expression for the reciprocal sum of the areas of the two triangles that is independent of this angle, but only of the interior angles of $\triangle DEF$ (and thus $\triangle XYZ$), and then computed the exact numerical values.

The scores for this problem are as follows:

Score	21	18	15	12	9	6	3	0
Count	84	20	8	46	38	9	48	355

The average score for this problem is 5.5 points, which is lower than expected. For students who received gold medals, the average score for this problem is 11.7 points. However, for those who received silver medals, the average score is only 3.3 points.

According to IMO's conventions for geometric problems, this problem did not provide a figure, requiring students to construct the corresponding geometric figures themselves. We noticed that many students, during the process of algebraically transforming the geometric problem, did not realize that their algebraic expressions relied on the relative positions of the points in the figure they had drawn on the paper. They assumed that the expressions held true for all situations, which is not a good habit. In Euclidean geometry, geometric expressions often differ due to the positive and negative variations caused by different point positions.

Transforming geometric conclusions into algebraic problems with proofs and providing complete geometric reasoning for those conclusions is the basis for credit according to IMO grading conventions. In other words, incomplete algebraic transformations alone cannot serve as the basis for partial credit.

3 Given positive integers m and n. Fix away to color the vertices of a regular $(2m+2n)$-gon so that $2m$ of them are black and the remaining $2n$ are white. Define the *coloring distance* $d(B, C)$ between two black points B, C to be the lesser of the number of white points on either side of line BC; define the *coloring distance* $d(W, X)$ between two white points W, X to be the lesser of the number of black points on either side of line WX.

A *black pairing scheme* $\mathcal{B}$ is to label all $2m$ black points as $B_1, \ldots, B_m, C_1, \ldots, C_m$, such that the m segments $B_i C_i (1 \leqslant i \leqslant m)$ do not intersect each other. For any black pairing scheme $\mathcal{B}$, denote

$$P(\mathcal{B}) = \sum_{i=1}^{m} d(B_i, C_i).$$

A *white pairing scheme* $\mathcal{W}$ is to label all $2n$ black points as $W_1, \ldots, W_m, X_1, \ldots, X_m$, such that the n segments $W_i X_i (1 \leqslant i \leqslant n)$ do not intersect each other. For any black pairing scheme $\mathcal{W}$, denote

$$P(\mathcal{W}) = \sum_{i=1}^{n} d(W_i, X_i).$$

Prove that, regardless of how the vertices are colored, there is always

$$\max_{\mathcal{B}} P(\mathcal{B}) = \max_{\mathcal{W}} P(\mathcal{W}),$$

where the maximum values on both sides of the equation are taken over all possible black pairing schemes $\mathcal{B}$ and white pairing schemes $\mathcal{W}$, respectively.

(Contributed by Xiao Liang)

Solution   Without loss of generality, place the $2m + 2n$ vertices on the unit circle. The Euclidean distances among these point vertices do not affect the discussion. We call the segments in the white pairing scheme *white segments*, and the segments in the black pairing scheme *black segments*.

Lemma 1   *For any black pairing scheme $\mathcal{B}$ and any white pairing scheme $\mathcal{W}$, the number of intersections $I(\mathcal{B}, \mathcal{W})$ of the $m + n$ segments is less than or equal to $P(\mathcal{B})$ and is also less than or equal to $P(\mathcal{W})$.* (For the black and white pairing scheme $\mathcal{B}$ and $\mathcal{W}$ shown in Fig. 3.3, the intersections are marked in gray, $I(\mathcal{B}, \mathcal{W}) = 5$, $P(\mathcal{B}) = 2 + 3 + 2 = 7$, $P(\mathcal{W}) = 1 + 3 + 1 = 5$.)

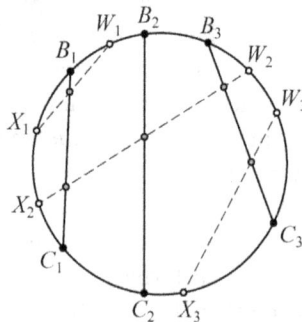

Fig. 3.3

Proof of Lemma 1 For any segment B_iC_i in $\mathcal{B}$, the number of white segments intersecting B_iC_i must not exceed $d(B_i, C_i)$ because each white segment corresponds to one white point on each side of segment B_iC_i. Taking the sum over all black segments in $\mathcal{B}$ gives the inequality $I(\mathcal{B}, \mathcal{W}) \leqslant P(\mathcal{B})$. Similarly, we can prove $I(\mathcal{B}, \mathcal{W}) \leqslant P(\mathcal{W})$. Hence, Lemma 1 is proved.

Lemma 2 *For any black pairing scheme $\mathcal{B}$, there always exists a white pairing scheme $\mathcal{W}$ such that $I(\mathcal{B}, \mathcal{W}) = P(\mathcal{B})$. Thus, $P(\mathcal{B}) = I(\mathcal{B}, \mathcal{W}) \leqslant P(\mathcal{W})$.*

We first explain how Lemma 2 proves the original problem. Applying Lemma 2 for all $\mathcal{B}$, we get

$$\max_{\mathcal{B}} P(\mathcal{B}) \leqslant \max_{\mathcal{W}} P(\mathcal{W}).$$

A symmetric discussion shows that

$$\max_{\mathcal{W}} P(\mathcal{W}) \leqslant \max_{\mathcal{B}} P(\mathcal{B}),$$

thus proving the problem.

Proof of Lemma 2 (Method 1) We prove the existence of white pairing scheme $\mathcal{W}$ on the number of white points $2n$: when $n = 0$, it is evident and does not need to be proved. Assume the existence has been proved for when there are $2n-2$ white points. Consider the case with $2n$ white points. Among all black point pairings B_iC_i, choose the one with the maximum $d(B_i, C_i)$, say B_1C_1. If $d(B_i, C_i) = 0$, any white pairing scheme will suffice. Now we assume that $d(B_i, C_i) > 0$. Let W_n be the first white point encountered when rotating clockwise starting from B_1, and let X_n be the first white point encountered when rotating clockwise starting from B_1. As shown in Fig. 3.4, consider the case where the two white points W_n and X_n are removed. Use $d'(B, C)$ to record the coloring distance in this case, and let

$$P'(\mathcal{B}) = \sum_{i=1}^{m} d'(B_i, C_i).$$

By inductive hypothesis, after removing W_n and X_n, there exists a white pairing scheme $\mathcal{W}'$ such that $I(\mathcal{B}, \mathcal{W}') = P'(\mathcal{B})$. We prove that adding segment W_nX_n to $\mathcal{W}'$ gives the required white pairing scheme $\mathcal{W}$.

We separate the discussion for each segment BC in $\mathcal{B}$. If BC intersects with W_nX_n, then removing W_nX_n is equivalent to removing one white point on each side of BC, so $d'(B, C) = d(B, C) - 1$. If BC does not intersect with W_nX_n, we want to show that $d(B, C) = d'(B, C)$. Note that if BC has more than $n + 1$ white points on the side of W_n and X_n, and

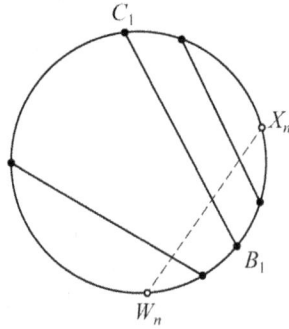

Fig. 3.4

the number of white points on the other side of BC does not exceed $n - 1$, then we must have $d(B, C) = d'(B, C)$. On the other hand, if BC has no more than n white points on the side of W_n and X_n, note that segments BC, $B_1 C_1$ and $W_n X_n$ do not intersect; therefore, the entire segment $B_1 C_1$ is on the side of BC with no more than n white points. So there must be $d(B, C) > d(B_1, C_1)$, which contradicts the maximality of $d(B_1, C_1)$. Hence, $d(B, C) = d'(B, C)$.

It can be seen that $P(\mathcal{B}) - P'(\mathcal{B})$ is exactly the number of black segments in $\mathcal{B}$ that intersect with $W_n X_n$, namely, $I(\mathcal{B}, \mathcal{W}) - I(\mathcal{B}, \mathcal{W}')$. Therefore, $I(\mathcal{B}, \mathcal{W}) = P(\mathcal{B})$. This completes the inductive proof.

Proof of Lemma 2 (Method 2) (This method is essentially the same as the previous proof but described differently.) For black segment BC in $\mathcal{B}$, we call the white points on the side with fewer white points an internal set of white points. (If $d(B, C) = n$, take the internal set of white points on either side.) Note that any two internal sets of white points (for two different black segments) either do not intersect or one is contained within the other. Take all the largest internal sets of white points, and treat any point not in any internal set of white points as its own internal set of white points. We only need to prove that there exists a pairing scheme for the white points such that no two white points corresponding to different white line segments are in the same internal set of white points. To this end, we first take the largest internal set of white points, connect its last white point with the next white point in clockwise direction, and remove these two white points, considering the remaining white points and corresponding internal sets of white points. Obviously, what is obtained in this way is a pairing scheme for the white points that meets the conditions because, by induction, it is easy to see that the number of white points in the largest internal set of white points at any time is less than or equal to the total number of white points in the remaining internal sets of white points.

Proof of Lemma 2 (Method 3) First, construct a weak white pairing scheme $\mathcal{W}'$, which pairs white points without requiring the corresponding white segments to be non-intersecting, such that $I(\mathcal{B}, \mathcal{W}') = P(\mathcal{B})$. In fact, if $W_1, W_2, \ldots, W_{2n}$ are the $2n$ white points sequentially arranged on the circumference of the circle, we only need to pair W_i with $W_{n+i}(1 \leqslant i \leqslant n)$ because this pairing will ensure that for any black segment, the white points on the side with fewer white points are connected to the other side.

Take all weak white pairing schemes $\mathcal{W}'$ that satisfy $I(\mathcal{B}, W') = P(\mathcal{B})$, and among them, let $\mathcal{W}_0$ be the one with the smallest total Euclidean distance of the white segments. We prove that n white segments in $\mathcal{W}'$ do not intersect each other. Otherwise, suppose the white segments WX and YZ intersect; without loss of generality, let them be sequentially W, Y, X, Z in clockwise direction on the circumference (as shown in Fig. 3.5). Since the black segments in $\mathcal{B}$ do not intersect, there cannot be a black segment connecting $\overparen{YX}$ to $\overparen{WZ}$ and a black segment connecting $\overparen{XZ}$ to $\overparen{WY}$ at the same time. We might as well suppose there is no black segment connecting $\overparen{YX}$ to $\overparen{WZ}$. We then replace YZ and WX with WY and ZX to obtain a weak white pairing scheme $\mathcal{W}_0'$, which clearly reduces the total Euclidean distance of all white segments.

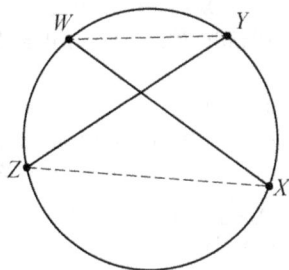

Fig. 3.5

Next, we compare $I(\mathcal{B}, \mathcal{W}_0)$ and $I(\mathcal{B}, \mathcal{W}_0')$:

- For any black segment connecting $\overparen{WY}$ to $\overparen{XY}$, or $\overparen{ZX}$ to $\overparen{XY}$, or $\overparen{WY}$ to $\overparen{WZ}$, or $\overparen{ZX}$ to $\overparen{WZ}$, the total number of intersections between them and WX, YZ and WY, ZX is 1.
- For any black segment connecting $\overparen{WY}$ to $\overparen{ZX}$, its intersection number with WY, WX, YZ, ZX is 1.

Therefore, $I(\mathcal{B}, \mathcal{W}_0) = I(\mathcal{B}, \mathcal{W}_0')$, which contradicts the fact that $\mathcal{W}_0$ minimizes the total Euclidean distance of all white segments. Hence, $\mathcal{W}_0$ is a valid white pairing scheme, completing the proof of Lemma 2.

Remark This problem exhibits a certain symmetry or, more precisely, a sense of 'duality,' that is, using white points and black points to calculate the coloring distance and get the same maximum value. This is a phenomenon that often occurs in mathematics. The key to solving the problem lies in the idea of counting twice, using the total number of intersections between black and white segments $I(\mathcal{B}, \mathcal{W})$ as a bridge to connect the coloring distances $P(\mathcal{B})$ and $P(\mathcal{W})$.

The scores for this problem are as follows:

Score	21	18	15	12	9	6	3	0
Count	36	11	8	10	36	6	31	470

The average score for this problem is 2.7 points, making it the lowest-scoring problem in this exam. For students who eventually entered the National Training Team, the average score for this problem reached 11.9 points, while those who did not enter the National Training Team but won gold medals had an average score of less than 4 points on this problem.

Second Day
(8:00–12:30; 30th December 2022)

④ Find the smallest integer $n \geqslant 3$ that satisfies the following: there exist n points $A_1, A_2, \ldots, A_n$ on a plane, with no three points being collinear, and for any $1 \leqslant i \leqslant n$, there exists $1 \leqslant j \leqslant n (j \neq i)$ such that segment $A_j A_{j+1}$ passes through the midpoint of segment $A_i A_{i+1}$. Here, $A_{n+1} = A_1$.

(Contributed by Fu Yunhao)

Solution First, we prove that $n = 6$ satisfies the condition of the problem. Let $P_1, P_2, \ldots, P_6$ be the vertices of a regular hexagon in clockwise order.

Let $A_1 = P_1, A_2 = P_3, A_3 = P_5, A_4 = P_2, A_5 = P_6, A_6 = P_4$, and then the 6 points satisfy the condition of the problem, as shown in Fig. 3.6.

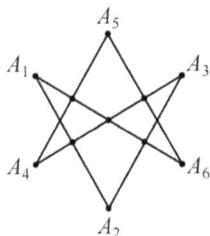

Fig. 3.6

In the following, we prove that $n \leqslant 5$ does not meet the condition. It is obvious that $n = 3$ does not meet the condition because for any $1 \leqslant i \leqslant n$, the described $1 \leqslant j \leqslant n$ needs to satisfy

$$j - i \not\equiv -1, 0, 1 \pmod{n}.$$

When $n = 4$, suppose A_1, A_2, A_3, A_4 satisfy the condition. Then A_1A_2 and A_3A_4 must bisect each other, and A_2A_3 and A_4A_1 must bisect each other, which is clearly impossible!

Thus, we only need to consider the case with $n = 5$. Suppose A_1, A_2, A_3, A_4, A_5 satisfy the condition. For any $1 \leqslant i \leqslant 5$, the described j needs to satisfy $j \equiv i + 2$, $i + 3 \pmod 5$. We discuss in the following two cases.

Case 1: Among segments $A_iA_{i+1}(i = 1, 2, 3, 4, 5)$, there are two segments bisecting each other. Due to symmetry, we might as well assume that A_1A_2, A_3A_4 bisect each other. Noting that an affine transformation of the plane does not affect the conclusion, we may assume that segments A_1A_2, A_3A_4 are perpendicular bisectors of each other and have the same length. Establish a plane rectangular coordinate system with their intersection point O as the origin, and set $A_1(0, -2)$, $A_2(0, 2)$, $A_3(2, 0)$, $A_4(-2, 0)$. Thus, the midpoint $M(1, 1)$ of A_2A_3 must lie on segment A_4A_5 or segment A_1A_5. By symmetry, we may assume that M lies on A_1A_5. Note that the slope of A_1M is 3, so we can set the coordinates of A_5 as $(1 + x, 1 + 3x)$, $x > 0$, as shown in Fig. 3.7.

Now we consider the midpoint N of A_4A_5. If it lies in the interior of segment A_1A_2, then $1 + x + (-2) = 0$, which gives $x = 1$. But now, A_4, A_2, A_5 are collinear, a contradiction. If N lies in the interior of segment A_2A_3,

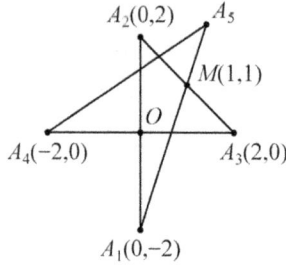

Fig. 3.7

then

$$\frac{1 + x + (-2)}{2} + \frac{1 + 3x}{2} = 2,$$

which also gives $x = 1$, a contradiction.

Case 2: No two segments in $A_i A_{i+1}(i = 1, 2, 3, 4, 5)$ bisect each other. By symmetry, we might as well assume that the midpoint B_1 of $A_1 A_2$ lies on $A_3 A_4$. Then the midpoint B_3 of $A_3 A_4$ lies on $A_5 A_1$, the midpoint B_5 of $A_5 A_1$ lies on $A_2 A_3$, the midpoint B_2 of $A_2 A_3$ lies on $A_4 A_5$, and the midpoint B_4 of $A_4 A_5$ lies on $A_1 A_2$. This implies that sequentially connecting A_1, A_2, A_3, A_4, A_5 will form a star (as shown in Fig. 3.8).

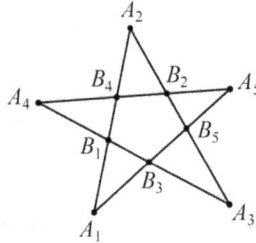

Fig. 3.8

By the law of sines, we have

$$1 = \prod_{i=1}^{5} \frac{\sin \angle A_i B_i B_{i+2}}{\sin \angle A_i B_{i+2} B_i} = \prod_{i=1}^{5} \frac{A_i B_{i+2}}{A_i B_i}$$

$$= \prod_{i=1}^{5} \frac{A_{i+1} B_{i+3}}{A_i B_i} < \prod_{i=1}^{5} \frac{A_{i+1} B_i}{A_i B_i} = 1,$$

where the indices are modulo 5. This is a contradiction!

To sum up, the desired minimum is $n = 6$.

Another proof of case 2 here is as follows. For $1 \leqslant i \leqslant 5$, consider the following equalities (since B_i is the midpoint of $A_i A_{i+1}$) and the inequality (the sum of two sides is greater than the third side):

$$B_i B_{i+3} + B_{i+3} A_{i+1} = B_i A_{i+1} = A_i B_i,$$

$$B_{i+3} A_{i+1} + B_{i+3} B_{i+1} > A_{i+1} B_{i+1}.$$

Summing over indices $i = 1, 2, 3, 4, 5$, the left and right sides of the two expressions sum up to be equal, but one is an equality and the other is a strict inequality, leading to a contradiction.

Remark Note that if the problem is changed to 'For any $1 \leqslant i \leqslant 5$, there exists $1 \leqslant j \leqslant n(j \neq i)$ such that line $A_j A_{j+1}$ passes through the midpoint of segment $A_i A_{i+1}$', the answer remains $n = 6$.

The proof for $n = 5$ is essentially the same as in Case 1 (but allows x to take negative values). For Case 2, we know that for the convex polygon formed by these five points, any two adjacent vertices cannot have adjacent indices, otherwise the line connecting them cannot pass through the midpoint of another segment. Therefore, the configuration must be as shown in Case 2.

This is a simple problem, and combinatorial geometry is a regular feature of IMO preliminary problems. However, simple graphs and finite extremes are not common in IMO preliminary problems, which makes it difficult to find similar problems.

Affine transformations, which can significantly reduce the complexity of subsequent discussions, were scarcely used by the candidates in the exam. This shows that there is still much room for improvement in broadening students' horizons in future teaching.

The most difficult part of this problem is to find the appropriate discussion method. The solution in the reference answer essentially only needs to discuss two cases, while many candidates have poor ability in handling classification discussions. Not only did some candidates discuss all 32 cases one by one, but there were also some solutions that used incorrect symmetry assumptions.

The scores for this problem are as follows:

Score	21	18	15	12	9	6	3	0
Count	295	82	118	39	38	4	12	20

The average score for this problem is 17.0 points. The average score for students who won the silver medal reached 17.9 points, which is more than

1 point lower than the average score for students who won gold medals but significantly higher than the average score for students who won the bronze medal (11.8 points). Although there is a slight barrier that caused a few students to score 0 points, the average score for this problem is similar to that of the first problem, and there are more students in the high-score range.

5. Prove that there exists a positive integer C such that the following statement holds:

For any infinite arithmetic sequence of positive integers $a_1, a_2, a_3, \ldots$, if the greatest common divisor of a_1 and a_2 is square-free, then there exists positive integer $m \leqslant C \cdot a_2^2$ such that a_m is square-free.

Note: A positive integer N is said to be square-free if it is not divisible by any square number greater than 1.

(Contributed by Qu Zhenhua)

Solution We prove that $C = 8$ satisfies the condition of the problem.

(1) First, consider the case where a_1 and a_2 are coprime. Let $d = a_2 - a_1$ be the common difference. For any prime p, if $p \mid d$, then $p \nmid a_1$, and thus p does not divide any a_n. If $p \nmid d$, then $p \nmid a_1$, and then any consecutive p^2 items in sequence $\{a_n\}$ form a complete system modulo p^2, where exactly one of them is divisible by p^2.

Let $N = 4a_2$. We prove that there exists $1 \leqslant n \leqslant N$ such that a_n is square-free. If $1 \leqslant m \leqslant N$ and a_m has square factors, then there exists prime p such that $p^2 \mid a_m$. Thus, $p \nmid d$ and $p \leqslant \sqrt{a_N}$. However, the number of terms in $a_1, a_2, \ldots, a_N$ divisible by p^2 does not exceed $\left\lceil \frac{N}{p^2} \right\rceil$. Therefore, assuming that the number of terms with square factors in $a_1, a_2, \ldots, a_N$ is M, then

$$M \leqslant \sum_{p \leqslant \sqrt{a_N}} \left\lceil \frac{N}{p^2} \right\rceil < \sum_{p \leqslant \sqrt{a_N}} \left(\frac{N}{p^2} + 1 \right)$$

$$\leqslant N \sum_{p \leqslant \sqrt{a_N}} \frac{1}{p^2} + \sqrt{a_N}$$

$$< \frac{N}{2} + \sqrt{a_N}.$$

The last inequality is due to

$$\sum_p \frac{1}{p^2} < \frac{1}{4} + \sum_{k=2}^{\infty} \frac{1}{(2k-1)^2}$$

$$< \frac{1}{4} + \sum_{k=2}^{\infty} \frac{1}{(2k-2)2k} = \frac{1}{2}.$$

When $N = 4a_2$, $\sqrt{a_N} < \sqrt{Na_2} = 2a_2 = \frac{N}{2}$, so $M < N$. Therefore, there exists a term in $a_1, a_2, \ldots, a_N$ that is square-free.

(2) Assume $gcd(a_1, a_2) = gcd(a_1, d) = q = q_1 \cdots q_l > 1$, where $q_1, \ldots, q_l$ are pairwise distinct primes. Note that each a_i is divisible by q. For each prime q_i, there exists a term a_{t_i} such that $v_{q_i}(a_{t_i}) = 1$. In fact, we can choose $t_i \in \{1, 2\}$ because a_1 and a_2 cannot both be divisible by q_i^2. By the Chinese remainder theorem, there exists $k \in \{1, 2, \ldots, q\}$ such that $k \equiv t_i \pmod{q_i}$, $1 \leqslant i \leqslant l$. Thus, for $1 \leqslant i$, there is $a_k \equiv a_{t_i} \pmod{q_i d}$, so $v_{q_i}(a_k) = 1$.

Consider the subsequence $a_k, a_{k+q}, a_{k+2q}, \ldots$ with common difference qd (divisible by q^2). It can be seen from the construction that $q_i^2 \nmid a_k (1 \leqslant i \leqslant l)$, that is, each term in the subsequence does not have $q_1, q_2, \ldots, q_l$ as square factors. Let $b_i = \frac{a_k + (i-1)q}{q}$. Then $b_1 < b_2 < \cdots$ is a positive integer arithmetic sequence with common difference d, and the first term $b_1 = \frac{a_k}{q}$ is coprime with d. By the conclusion in (1), there exists $i \leqslant 4b_2$ such that b_i has no square factors. And since q is coprime with b_i, $a_{k+(i-1)q} = qb_i$ is also square-free. Therefore,

$$k + (i-1)q \leqslant iq \leq 4b_2 q = 4a_{k+q} < 4(k+q)a_2 \leqslant 8da_2 < 8a_2^2.$$

Remark The proof of this problem requires estimating the number of terms containing square factors from $a_1, a_2, \ldots, a_N$ (or from a subsequence), denoted as $S(N)$, and then proving the existence of a term that is square-free by showing $S(N) < N$. Note that primes can be classified into three categories:

- Irrelevant prime p: For any n, $p^2 \nmid a_n$.
- Good prime p: Among consecutive p^2 terms in a_n, there is exactly one term that is divisible by p^2.
- Bad prime p: Among consecutive p terms in a_n, there is exactly one term that is divisible by p^2. (This is equivalent to $p|a_1$, $p|d = a_2 - a_1$, $p^2 \nmid d$.)

There are only a finite number of bad primes, denoted as $q_1, \ldots, q_t$. Let $Q = q_1 \cdots q_t$, and then $Q \mid d$. In the reference solution, we first consider the case without bad primes, where $S(N) < \frac{N}{2} + \sqrt{a_N}$. For the general case, select a subsequence such that it does not contain bad prime factors. Of course, this can also be written in one step by selecting a subsequence first and then estimating the number of terms with square factors. Many students adopt this approach in their exam papers, which is essentially the same as the reference solution.

In addition, bad primes and good primes can be discussed together. Suppose $N = 5a_2^2 - u \geqslant 4a_2^2, 0 \leqslant u < Q$ such that $Q \mid N$. Among consecutive Q terms in a_n, there are $\varphi(Q)$ terms that are not divisible by any q_i^2. Hence, among $a_1, a_2, \ldots, a_N$, there are $\frac{N\varphi(Q)}{Q}$ terms that are not divisible by any q_i^2. Denote

$$[N]_Q = 1 \leqslant n \leqslant N : q_i^2 \nmid a_n, \quad \forall 1 \leqslant i \leqslant t.$$

$[N]_Q$ is the union of $\varphi(Q)$ congruence classes modulo Q. Suppose p is a good prime. Remove n such that $p^2 \mid a_n$ from $p^2 \mid a_n$. Hence, one term of every p^2 term is removed in each congruence class modulo Q, and no more than $\varphi(Q)\left\lceil \frac{N}{Qp^2} \right\rceil$ terms are removed. Note that if $p^2 \mid a_n$, then $p^2 \mid \frac{a_n}{Q}$, so $p \leq \sqrt{\frac{a_N}{Q}}$. Therefore,

$$N - S(N) \geqslant |[N]_Q| - \sum_{p \leq \sqrt{\frac{a_N}{Q}}} \varphi(Q) \left\lceil \frac{N}{Qp^2} \right\rceil$$

$$\geqslant \frac{N\varphi(Q)}{Q} \left(1 - \sum_p \frac{1}{p^2} \right) - \varphi(Q) \sqrt{\frac{a_N}{Q}}$$

$$> \frac{N\varphi(Q)}{2Q} - \varphi(Q) \sqrt{\frac{a_N}{Q}}$$

$$\geqslant \frac{N\varphi(Q)}{2Q} - \varphi(Q) \sqrt{\frac{Na_2}{Q}}$$

$$= \varphi(Q) \sqrt{\frac{N}{Q}} \left(\frac{1}{2} \sqrt{\frac{N}{Q}} - \sqrt{a_2} \right)$$

$$\geqslant \varphi(Q) \sqrt{\frac{N}{Q}} \left(\frac{1}{2} \sqrt{\frac{4a_2^2}{Q}} - \sqrt{a_2} \right) > 0.$$

Consequently, $C = 5$ also satisfies the requirement.

The scores for this problem are as follows:

Score	21	18	15	12	9	6	3	0
Count	143	13	18	2	25	36	78	293

The average score for this problem is 6.9 points. The average score for students who won gold medals is 15.5 points, while the average score for students who won silver and bronze medals is only 2.6 points.

6 There are $n(n \geqslant 8)$ airports, and there are direct one-way flights between some airports. For any two airports a and b, there is at least one direct one-way flight either from a to b or from b to a (there may be flights both from a to b and from b to a). It is known that for any set $A(1 \leqslant |A| \leqslant n-1)$ consisting of several airports, there are at least $4 \cdot \min\{|A|, n - |A|\}$ one-way flights from airports in A to airports outside A.

Prove that for any airport x, it is possible to depart from x and return to x via no more than $\sqrt{2n}$ direct one-way flights.

(Contributed by Ai Yinghua)

Solution   Represent each airport as a point. If there is a direct one-way flight from airport a to airport b, draw a directed edge $\overrightarrow{ab}$, thus obtaining a directed graph G. Denote $r = \lfloor \sqrt{\frac{n}{2}} \rfloor$, and then $r \geqslant 2$. We prove that for any vertex v, there exists a directed cycle of length not exceeding $2r$ that passes through v. This will establish the proposition stated in the problem.

We use proof by contradiction. Assume that there does not exist a directed cycle of length not exceeding $2r$ passing through v in G. For each positive integer $k \leqslant r$, define

$N_k = \{x \in V(G) \mid x \neq v$, and the length of the shortest directed path form v to x is $k\}$.

Let $T_k = N_1 \bigcup \cdots \bigcup N_k$. First, we prove the following fact:

(*) *For* $1 \leqslant k \leqslant r-1$, *if the directed edge* $\overrightarrow{xy}$ *is from* T_k *to its complement* T_k^c, *then there must be* $x \in N_k$ *and* $y \in N_{k+1}$.

The proof of fact (*) is as follows: It is obvious that $y \neq v$; otherwise, from v to x there is a directed path of length not exceeding k, and combing edge $\overrightarrow{xy} = \overrightarrow{xu}$, we get a directed cycle passing through v with length not exceeding $1 + k \leqslant r < 2r$, a contradiction. Suppose $x \in N_i(1 \leqslant i \leqslant k)$, and then from v to x there is a directed path P with length i. Combining P with $\overrightarrow{xy}$, we get a directed edge from v to y with length not exceeding $i+1$. Let d be the shortest path from v to y, and then $d \leqslant i+1 \leqslant k+1$. From $y \in (T_k \cup \{v\})^c$, we know $d \geqslant k+1$, and thus $d = k+1$, $i = k$. Therefore, $x \in N_k$, $y \in N_{k+1}$.

In addition, we assert $|T_r| \geqslant \frac{n}{2}$.

Proof of the Assertion   We use proof by contradiction. Assume $|T_r| < \frac{n}{2}$. For each $1 \leq i \leqslant r-1$, consider the set M of all directed edges from T_i to T_i^c. On one hand, by the conditions of the problem, we have

$$|M| \geqslant 4 \cdot \min |T_i|, \quad n - |T_i| = 4 \cdot |T_i|.$$

On the other hand, from fact $(*)$ we know that for any edge $\overrightarrow{xy}$ in M, there is $x \in N_i$, $y \in N_{i+1}$. Then it follows that the number of edges in M does not exceed the number of ordered pairs $(x, y) \in N_i \times N_{i+1}$, that is, $|M| \leqslant |N_i| \cdot |N_{i+1}|$. Combining these estimates, we obtain for any $1 \leqslant i \leqslant r-1$, there is

$$4 \cdot |T_i| \leqslant |N_i| \cdot |N_{i+1}|. \qquad \qquad ①$$

Denote $t_i = |T_i| = |N_1| + \cdots + |N_i|$. From the problem's conditions, we know $t_1 = |N_1| \geqslant 4 = 2^2$. From fact $(*)$ we know that all directed edges $\overrightarrow{xy}$ from T_1 to its complement T_1^c satisfy $y \in N_2$. In particular, $|N_2| \geqslant 1$. Thus, $|N_1| \leqslant |T_2| - 1 \leqslant |T_r| - 1 < \frac{n}{2} - 1$. Similar to fact $(*)$, we can prove that all directed edges $\overrightarrow{xy}$ from $\{v\} \cup N_1$ to its complement satisfy $x \in N_1$, $y \in N_2$. Therefore,

$$|N_1| \cdot |N_2| \geqslant 4 \cdot \min\{1 + |N_1|, \quad n - 1 - |N_1|\} = 4(1 + |N_1|).$$

Then it follows that

$$t_2 = |N_1| + |N_2| \geqslant |N_1| + \frac{4(1 + |N_1|)}{|N_1|} > |N_1| + 4 \geqslant 8,$$

and hence $t_2 \geqslant 9 = 3^2$.

Based on this, we use induction to prove that for $1 \leqslant k \leqslant r$, there is $t_k \geqslant (k+1)^2$. Assume that for a certain $3 \leqslant m \leqslant r$, $t_{m-2} \geqslant (m-1)^2$ and $t_{m-1} \geqslant m^2$ have already been established. From ① we know that for

$1 \leqslant i \leqslant r - 1$, there is $4t_i \leq (t_i - t_{i-1})(t_{i+1} - t_i)$, and thus

$$t_m \geqslant t_{m-1} + \frac{4t_{m-1}}{t_{m-1} - t_{m-2}}$$

$$= t_{m-2} + (t_{m-1} - t_{m-2}) + \frac{4t_{m-1}}{t_{m-1} - t_{m-2}}$$

$$\geqslant t_{m-2} + 2\sqrt{4t_{m-1}}$$

$$\geqslant (m-1)^2 + 2\sqrt{4m^2}$$

$$= (m+1)^2,$$

which completes the induction proof. Thus, $|T_r| \geqslant (r+1)^2 = \left(\lfloor\sqrt{\frac{n}{2}}\rfloor + 1\right)^2 > \frac{n}{2}$ contradicts the previous assumption $|T_r| < \frac{n}{2}$. Therefore, we have completed the proof of the assertion.

Similarly, define $K_i = \{x \in V(G) \mid x \neq v,$ and the length of the shortest directed path form x to v is $i\}$, and let $U_i = K_1 \cup \cdots \cup K_i$. Symmetrically, we can prove that $|U_r| \geqslant \frac{n}{2}$.

Note that T_r and U_r are subsets of $V(G)\backslash\{v\}$, and there is $|T_r| \geqslant \frac{n}{2}$, $|U_r| \geqslant \frac{n}{2}$, so T_r and U_r must intersect. Take $x \in T_r \cap U_r$. Then from v to x there is a directed path of length not exceeding r, and from x to v there is a directed path of length not exceeding r. The union of the two paths forms a directed cycle passing through v of length not exceeding $2r$, which is a contradiction!

Remark We can also estimate the size of $S_i = \{v\} \cup T_i$ using the conditions of the problem and prove the conclusion of the problem by establishing $|S_r| \geqslant \frac{n}{2} + 1$. It is easy to see that all directed edges $\overrightarrow{xy}$ from S_i to S_i^c satisfy $x \in N_i$, $y \in N_{i+1}$, so there is $|N_i| \cdot |N_{i+1}| \geqslant 4 \cdot \min\{|S_i|, n - |S_i|\}$. Under the hypothesis of $|S_r| < \frac{n}{2} + 1$ in the proof by contradiction, note that all directed edges $\overrightarrow{xy}$ from S_{r-1} to its complement satisfy $y \in N_r$. Thus, we can get $|N_r| \geqslant 1$, $|S_{r-1}| < \frac{n}{2}$. Therefore, for $1 \leqslant i \leq r - 1$, there is $|N_i| \cdot |N_{i+1}| \geqslant 4 \cdot |S_i|$. From this, we can use induction to prove that for $k \leq r - 1$, there is $|S_k| \geqslant 1 + (1+k)^2$.

The scores for this problem are as follows:

Score	21	18	15	12	9	6	3	0
Count	70	8	21	7	34	6	22	440

The average score for this problem is 4.0 points. The average score for students who entered the National Training Team is close to 18 points. The average score for other students who won gold medals on this problem is less than 7 points.

Analysis of the Scores

Score Distribution

The number of candidates participating in the final this year increased compared to last year, with a total of 608 candidates. The average score for the entire exam paper (with a full score of 126 points) was $m = 53.4$, with a standard deviation of $\sigma = 24.4$.

Gold and Silver Medal Cutoff Scores

The theoretical number of silver medalists is 40% of all candidates. In practice, 267 candidates (43.9%) received silver medals, with the cutoff score being 36 points.

Within the silver medal range ± 21 points (i.e., $[15, 57]$ points), the correlation between the total score and the scores of individual problems is shown in the following table:

Problem Number	Problem 1	Problem 2	Problem 3	Problem 4	Problem 5	Problem 6
Correlation with Total Score	0.49	0.39	0.20	0.51	0.40	0.25

The theoretical number of gold medalists is 30% of all candidates. In practice, 204 students (33.6%) received gold medals, with the cutoff score being 63 points.

Within the gold medal range ± 21 points (i.e., $[42, 84]$ points), the correlation between the total score and the scores of individual problems is shown in the following table:

Problem Number	Problem 1	Problem 2	Problem 3	Problem 4	Problem 5	Problem 6
Correlation with Total Score	0.00	0.42	0.23	0.01	0.63	0.41

China National Training Team for the International Mathematical Olympiad

For the two-day test, the 60th place score on the first day is 42, and the 60th place score on the second day is 51 (the 46th–97th place scores on the first day are all 42, and the 59th–75th place scores on the second day are all 51).

The lowest score required to be selected for the 2023 China National Training Team for the International Mathematical Olympiad is 90 points.[2] The distribution of their scores over the two days is as follows:

Score	First-day Candidates	Second-day Candidates
63	6	21
60	3	8
57	5	5
54	5	4
51	8	9
48	5	3
45	3	1
42	11	6
39	4	1
36	2	0
33	4	2
30	2	0
27	2	0

Within the range of ± 21 points from the National Training Team cutoff (i.e., [69, 111] points), the correlation between total score and individual problem scores is as follows:

Problem Number	Problem 1	Problem 2	Problem 3	Problem 4	Problem 5	Problem 6
Correlation with Total Score	0.05	0.29	0.25	0.10	0.18	0.46

---

[2]For the students with a total score of 90 (a total of 13 students), ranking was done according to the pre-announced team selection rules: 'If the total scores are the same, priority is given based on weighted scores calculated as follows: Weighted Total Score = (Problem 1 Score + Problem 4 Score) + (Problem 2 Score + Problem 5 Score) × 1.1 + (Problem 3 Score + Problem 6 Score) × 1.2. The higher weighted total score ranks first. If the weighted total scores are still the same, the scores of Problem 6, Problem 3, Problem 5, Problem 2, Problem 4, and Problem 1 are compared in sequence, and the higher score ranks first.' In this competition, only the 'weighted total score' ranking was used.

4

The 21st Chinese Girls' Mathematical Olympiad

The 21st Chinese Girls' Mathematical Olympiad (CGMO) was held from 10th to 14th August 2022, at Tianjin Nankai High School. The competition was hosted by the Chinese Mathematical Society, and organized by the Tianjin Mathematical Society and Tianjin Nankai High School.

A total of 143 female students from 36 teams, representing various provinces, municipalities, autonomous regions, the Hong Kong Special Administrative Region of China, and Macau Special Administrative Region of China, participated in the competition.

After two rounds of contests (4 hours and four questions each), Yuan Lai from The High School Affiliated with Renmin University of China and 43 other students won the gold medals (first prize), Wang Kexin from Wuhan No. 2 High School in Hubei Province and 53 other students won the silver medals (second prize), and Qi Yufei from The High School Attached to Northwest Normal University and 44 other students won the bronze medals (third prize). In addition, Yuan Lai and 17 other students (tied for 15th place) qualified to participate in the 2022 China Mathematical Olympiad (CMO), and the Beijing team won the first place in the team.

Director of the main examination committee: Xiong Bin (East China Normal University).

Members of the main examination committee (in alphabetical order):

First Day
(8:00–12:30; 12th August 2022)

1 Consider all the real sequences $x_0, x_1, x_2, \ldots, x_{100}$ satisfying the following conditions:

(1) $x_0 = 0$;

(2) For any integer $i, 1 \leqslant i \leqslant 100$, $1 \leqslant x_i - x_{i-1} \leqslant 2$ holds.

Find the largest positive integer $k \leqslant 100$ such that

$$x_k + x_{k+1} + \cdots + x_{100} \geqslant x_0 + x_1 + \cdots + x_{k-1}$$

holds for any such sequence $x_0, x_1, x_2, \ldots, x_{100}$.

(Contributed by Zhang Sihui)

Solution On one hand, if $x_i = 2i, 1 \leqslant i \leqslant 34$, $x_{34+j} = x_{34} + j = 68 + j$, $1 \leqslant j \leqslant 66$, then the sequence satisfies the given conditions. We have

$$\sum_{j=68}^{100} x_j - \sum_{i=0}^{67} x_i = \sum_{j=1}^{33} (x_{67+j} - x_{34+j}) - \sum_{i=1}^{34} x_i = 33^2 - 34 \times 35 < 0.$$

This example shows that for $k \geqslant 68$, the condition is not satisfied.

On the other hand, for any sequence $x_0, x_1, x_2, \ldots, x_{100}$ that satisfies the given conditions, it is easy to see that $x_i \leqslant 2i, 1 \leqslant i \leqslant 100$.

For $0 \leqslant s < t \leqslant 100$, the inequality $x_t - x_s \geqslant t - s$ holds. Thus,

$$\sum_{j=67}^{100} x_j - \sum_{i=0}^{66} x_i = \sum_{j=1}^{34} (x_{66+j} - x_{32+j}) - \sum_{i=1}^{32} x_i$$

$$\geqslant 34^2 - \sum_{i=1}^{32} 2i$$

$$= 34^2 - 32 \times 33$$

$$> 0.$$

Based on the above analysis, the desired largest k is 67.

2 Let n be a positive integer. There are $3n$ women's volleyball teams in the tournament. Each two team plays at most once (there are no ties in volleyball games). Assume that a total number of $3n^2$ games have been played. Prove that there exists a team whose number of wins and number of losses are both greater than or equal to $\frac{n}{4}$.

(Contributed by Ai Yinghua)

Solution Assume that the conclusion does not hold. Let the number of wins for teams $P_1, \ldots, P_k$ be less than $\frac{n}{4}$, and the number of wins for the remaining $3n - k$ teams is not less than $\frac{n}{4}$. By assumption, we know that the number of losses is less than $\frac{n}{4}$. Since the total number of wins for all teams is $3n^2$, there must be a team with at least n wins, and hence $3n - k > 0$.

The number of games among $P_1, \ldots, P_k$ does not exceed the sum of their wins, so it does not exceed $\frac{nk}{4}$. Similarly, the number of games among $Q_1, \ldots, Q_{3n-k}$ does not exceed the sum of their losses, so it is less than $\frac{n(3n-k)}{4}$ ($3n - k > 0$ is used here).

The number of games between teams $P_1, \ldots, P_k$ and $Q_1, \ldots, Q_{3n-k}$ is at most $k(3n - k)$, so it can be seen that the total number of games N satisfies

$$N < \frac{nk}{4} + \frac{n(3n - k)}{4} + k(3n - k) \leqslant \frac{3n^2}{4} + \frac{9n^2}{4} = 3n^2,$$

which contradicts the conditions. Therefore, the assumption is false, and the original statement is proved.

3 Let I be the incenter of a triangle ABC with $AB > AC$, and let AM be the median. The line passing through I and perpendicular to

BC meets line AM at point L. Let J be the reflection of I about A. Prove that $\angle ABJ = \angle LBI$.

(Contributed by Luo Zhenhua)

Solution   We need the following lemma.

Lemma   *In* $\triangle ABC$, *the incenter is* I, *and the incircle* $\odot I$ *touches sides* BC, CA, AB *at points* A_1, B_1, C_1, *respectively. Suppose* $A_1 I$ *intersects* $B_1 C_1$ *at point* L, *and then* AL *passes through the midpoint* M *of* BC.

Proof of the lemma   As shown in Fig. 4.1, draw a line passing through L and parallel to BC, intersecting AC at point X and AB at point Y. Connect IB_1, IC_1, IX, IY.

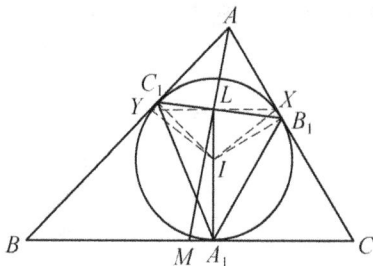

Fig. 4.1

From $A_1 L \perp BC, XY \parallel BC$, it follows that $A_1 L \perp XY$. Thus, $\angle ILX = \angle IB_1 X = 90°$. Therefore, points I, L, X, B are concyclic. Likewise, points I, L, C_1, Y are concyclic. Hence, $\angle IXB_1 = \angle ILB_1 = \angle IYC_1$.

Combining $\angle IB_1 X = \angle IC_1 Y = 90°, IB_1 = IC_1$, we have $\triangle IB_1 X \cong \triangle IC_1 Y$. Thus, $IX = IY$.

Also note that $IL \perp XY$, so L is the midpoint of XY. Combining $XY \parallel BC$, it can be seen that AL passes through the midpoint M of BC. The lemma is proved.

We return to the original problem. As shown in Fig. 4.2, let the incircle $\odot I$ of $\triangle ABC$ touch sides BC, CA, AB at points A_1, B_1, C_1, respectively. By the lemma, point L lies on $B_1 C_1$. Suppose S is the midpoint of BI and T is a point on AS, satisfying $\angle TBS = \angle BAS$.

In the following we will prove that points B, T, L are collinear.

From $\angle TBS = \angle BAS$, we know $\triangle BST \backsim \triangle ASB$, and hence $BS^2 = SA \cdot ST$. And since $BS = SI$, $IS^2 = SA \cdot ST$. Therefore, $\triangle IST \backsim \triangle ASI$,

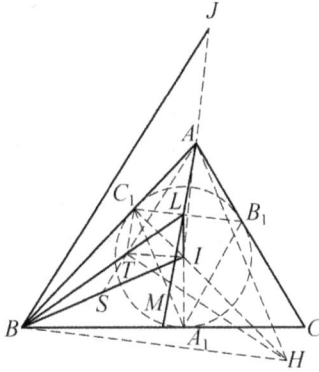

Fig. 4.2

so

$$\angle BTI = \angle BTS + \angle ITS = \angle ABI + \angle AIB = 180° - \angle BAI.$$

Let H be the orthocenter of $\triangle AIB$. According to the properties of the orthocenter, we have $\angle BHI = \angle BAI$, and hence points B, T, I, H are concyclic. Therefore, $\angle STH = \angle BTH - \angle BTS = \angle BIH - \angle ABI = 90°$. Combining $\angle AC_1 H = 90°$, it follows that points A, C_1, T, H are concyclic, and AH is the diameter of this circle.

Consequently,

$$\angle ITC_1 = \angle HTC_1 - \angle HTI = 180° - \angle HAC_1 - \angle HBI$$

$$= 180° - (90° - \angle ABI) - (90° - \angle ABI - \angle BAI)$$

$$= 2\angle ABI + \angle BAI.$$

And by

$$\angle ILB_1 = \angle IC_1 L + \angle LIC_1 = \frac{1}{2}\angle BAC + \angle ABA_1 = \angle BAI + 2\angle ABI,$$

we have $\angle ITC_1 = \angle ILB_1$, and hence points I, T, C_1, L are concyclic. Then we get $\angle ITL = \angle IC_1 L = \angle BAI$, and thus $\angle BTI + \angle ITL = 180°$. Therefore, points B, T, L are collinear, and $\angle LBI = \angle TBI = \angle BAS$.

And since AS is the midline of $\triangle BIJ$, $AS \parallel BJ$, $\angle BAS = \angle ABJ$. Therefore, $\angle ABJ = \angle LBI$, completing the proof.

4 Given a prime $p \geqslant 5$. Find the number of distinct residues modulo p of the product of three consecutive positive integers.

(Contributed by Wang Bin)

Solution   Denote the set $D = \{0, 1, \ldots, p-1\}$ and the polynomial

$$f(x) = (x-1)x(x+1) = x^3 - x.$$

The congruence and the congruence symbol $\equiv$ in this problem both represent congruence modulo p. For $k = 0, 1, 2, 3$, denote

$$B_k = \{b \in D | \text{there are } k \text{ elements } a \in D \text{ such that } f(a) \equiv b\}.$$

Consider element b in B_2, and there exists $a_1 \neq a_2$ such that $f(a_1) \equiv f(a_2) \equiv b$. Thus,

$$0 \equiv (a_1^3 - a_1) - (a_2^3 - a_2) = (a_1 - a_2)(a_1^2 + a_1 a_2 + a_2^2 - 1)$$

$$\Rightarrow a_1^2 + a_1 a_2 + a_2^2 \equiv 1.$$

In this way, for $a_3 = -a_1 - a_2$, there is

$$a_3^2 + a_3 a_1 + a_1^2 = (a_1 + a_2)^2 - (a_1 + a_2)a_1 + a_1^2 \equiv 1,$$

which implies $f(a_3) \equiv f(a_1) \equiv b$. By the definition of B_2, we know that a_3 is congruent to one of a_1 and a_2, so $a_3 = -2a_1$, or $a_3 = -2a_2$. We may as well set the former, and then $a_1^2 + a_1 a_2 + a_2^2 \equiv 3a_1^2 \equiv 1$, $3b \equiv 3a_1^3 - 3a_1 \equiv -2a_1$. When 3 is a quadratic residue modulo p, there are two of a_1; when 3 is not a quadratic residue modulo p, a_1 does not exist. Therefore, using Legendre symbol, there are exactly $1 + \left(\frac{3}{p}\right)$ of a_1, and the number of the corresponding b is also exactly $1 + \left(\frac{3}{p}\right)$. Thus, $|B_2| = 1 + \left(\frac{3}{p}\right)$.

It is obvious that $|B_1| + 2|B_2| + 3|B_3| = |D| = p$.

On the other hand, we consider (u, v) satisfying $f(u) \equiv f(v)$ and $u \neq v$ (called a collision), which is equivalent to $u^2 + uv + v^2 \equiv 1$ and $u - v \neq 0$. After the substitution of $(x, y) \equiv \left(\frac{u-v}{2}, \frac{u+v}{2}\right)$ (that is, $(u, v) \equiv (x+y, y-x)$, such that (x, y) and (u, v) are in one-to-one correspondence), it transforms to $x^2 + 3y^2 \equiv 1$ and $x \neq 0$.

Now, we consider the number of (x, y) satisfying $x^2 + 3y^2 \equiv 1$ (i.e., $(3y)^2 \equiv 3 - 3x^2$) modulo p:

$$T = \sum_{x=1}^{p} 1 + \left(\frac{3 - 3x^2}{p}\right) = p + \left(\frac{-3}{p}\right) \sum_{x=1}^{p} \left(\frac{x^2 - 1}{p}\right)$$

$$= p + \left(\frac{-3}{p}\right) \sum_{x=2}^{p} \left(\frac{\frac{x+1}{x-1}}{p}\right) = p + \left(\frac{-3}{p}\right) \sum_{x=1=1}^{p-1} \left(\frac{1 + \frac{2}{x-1}}{p}\right)$$

$$= p + \left(\frac{-3}{p}\right) \sum_{z=1}^{p-1} \left(\frac{z+1}{p}\right) = p - \left(\frac{-3}{p}\right).$$

The above equation has used the reduced residue system modulo p of $z = \frac{2}{x-1}$ and $\sum_{k-1}^{p} \left(\frac{k}{p} \right) = 0$. Among these T pairs of (x, y), there are exactly $1 + \left(\frac{3}{p} \right)$ pairs where $x \equiv 0$ (i.e., $3y^2 \equiv 1$), so the number of colliding ordered pairs (u, v) is

$$M = T - 1 - \left(\frac{3}{p} \right) = p - \left(\frac{-3}{p} \right) - 1 - \left(\frac{3}{p} \right) = T - |B_2|.$$

There are exactly $\frac{M}{2}$ such colliding unordered pairs (u, v). Since the elements in B_3 correspond to 3 collisions, the elements in B_2 correspond to 1 collision, and the elements in B_1 and B_0 correspond to 0 collision, $|B_2| + 3|B_3| = \frac{M}{2}$.

Thus, $|B_3| = \frac{M - 2|B_2|}{6}$. Therefore, the number of all possible residues for $f(x)$ modulo p is

$$|B_1| + |B_2| + |B_3| = p - |B_2| - 2|B_3| = p - \frac{M + |B_2|}{3}$$

$$= p - \frac{T}{3} = \frac{2p + \left(\frac{-3}{p} \right)}{3} = \left\lfloor \frac{2p + 1}{3} \right\rfloor.$$

Remark 1 Another approach to calculating the number T is as follows:

Consider the congruence equation $x^2 + 3y^2 \equiv z^2$ equivalent to $3y^2 \equiv z^2 - x^2 = (z + x)(z - x)$ with a total of p^2 solutions ($y \equiv 0$ has exactly $2p - 1$ solutions, $y \equiv 1$ has exactly $p - 1$ solutions, etc.). Among these, the number of solutions for $z \equiv 0$ is

$$m_0 = 1 + (p - 1) \left[1 + \left(\frac{-3}{p} \right) \right],$$

the rest of the number of solutions of $z \equiv 1, z \equiv 2, \ldots, z \equiv p - 1$ are equal (one can use $(x, y, z) \leftrightarrow (kx, ky, kz)$ to match the solutions of $z \equiv 1$ to the solutions of $z \equiv k$), and all are $T = \frac{p^2 - m_0}{p - 1} = p - \left(\frac{-3}{p} \right)$.

If you are not familiar with the Legendre symbol, you can first get $0 \leqslant m_0 \leqslant 2p$, and then you can derive the number $T \in \{p - 1, p, p + 1\}$ for $x^2 + 3y^2 \equiv 1$. Then, it follows that the solution to this problem is $p - \frac{T}{3}$, which is an integer.

Remark 2 Another way to understand the solution number $T = p - \left(\frac{-3}{p} \right)$: for each y, the number of x that satisfy $x^2 \equiv 1 - 3y^2$ is $1 + \left(\frac{1 - 3y^2}{p} \right)$.

Therefore,

$$T - p = \sum_{y=0}^{p-1} \left(\frac{1 - 3y^2}{p} \right) \equiv \sum_{y=0}^{p-1} (1 - 3y^2)^{\frac{p-1}{2}}$$

$$= \sum_{y=0}^{p-1} \left((-3)^{\frac{p-1}{2}} y^{p-1} + \cdots + 1 \right) \equiv (-3)^{\frac{p-1}{2}} (p - 1) = -\left(\frac{-3}{p} \right).$$

Second Day
(8:00–12:30; 13th August 2022)

5   As shown in Fig. 4.3, points K and L are in the interior of triangle ABC, point D lies on side AB. It is known that points B, K, L, C are concyclic, and $\angle AKD = \angle BCK, ALD = \angle BCL$. Prove that $AK = AL$.

(Contributed by He Yijie)

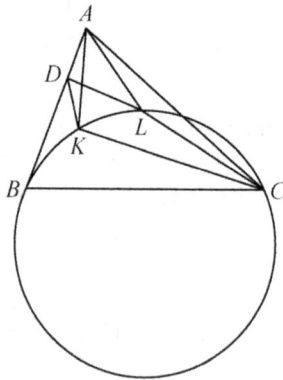

Fig. 4.3

Solution   As shown in Fig. 4.4, let the extensions of AK and AL intersect the circle passing through points B, K, L and C at points X and Y, respectively. Connect BX and BY. By combining the given conditions, we get $\angle AKD = \angle BCK = \angle BXK, \angle ALD = \angle BCL = \angle BYL$. Therefore, $DK \parallel BX, DL \parallel BY$. Thus, we have

$$\frac{AK}{AX} = \frac{AD}{AB} = \frac{AL}{AY}.$$

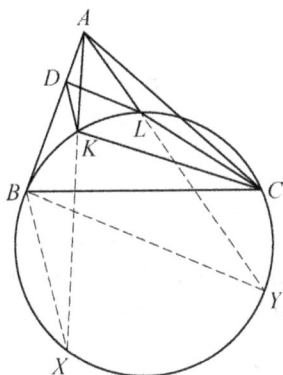

Fig. 4.4

Since K, L, Y, X are concyclic, by the power of a point theorem, there is

$$AK \cdot AX = AL \cdot AY.$$

Multiplying the above two equalities gives $AK^2 = AL^2$, namely, $AK = AL$.

6 Find all positive integers n with the following property: there exist nonempty finite sets of integers A, B, such that for any integer m, exactly one of the following three statements is true:

(i) there exists $a \in A$, such that $m \equiv a \pmod{n}$;
(ii) there exists $b \in B$, such that $m \equiv b \pmod{n}$;
(iii) there exist $a \in A$ and $b \in B$, such that $m \equiv a + b \pmod{n}$.

(Contributed by Fu Yunhao)

Solution Let $A + B = \{a + b \mid a \in A, b \in B\}$. The question can be understood as: the remainders of A modulo n, the remainders of B modulo n, and the remainders of $A + B$ modulo n exactly constitute a division of all the remainders modulo n.

(1) If $n > 1$ is odd, let $n = 2k + 1, k \in \mathbf{Z}_{>0}$. Taking $A = \{k\}$, $B = \{k + 1, k + 2, \ldots, 2k\}$, then $A + B = \{2k + 1, 2k + 2, \ldots, 3k\}$. Thus, $A, B, A + B$ are disjoint, and their union is exactly $2k + 1$ consecutive numbers, forming a complete residue system modulo $2k+1$ that satisfies the conditions.

(2) If $n > 1$ satisfies the conditions, then for any integer $d > 1$, dn also satisfies the conditions. In fact, let the corresponding sets for n be A

and B. Let

$$A' = \{a + xn \mid a \in A, x = 0, 1, \ldots, d - 1\},$$
$$B' = \{b + xn \mid b \in B, x = 0, 1, \ldots, d - 1\}.$$

We verify that A' and B' satisfy the requirements for dn.

Since the remainders of $A', B', A' + B'$ modulo n are exactly the remainders of $A, B, A + B$ modulo n, it follows that the remainders of $A', B', A' + B'$ modulo n are disjoint, and of course the remainders of $A', B', A' + B'$ modulo dn are disjoint, so at most one of (i), (ii), (iii) holds.

Consider any integer m. If there exists $a \in A$ such that $m \equiv a \pmod{n}$, then m is congruent to one of $a + xn (x = 0, 1, \ldots, d - 1)$ modulo dn, so (i) holds. Similarly, if there exists $b \in B$ such that $m \equiv b \pmod{n}$, then (ii) holds. If there exist $a \in A$ and $b \in B$ such that $m \equiv a + b \pmod{n}$, then m is congruent to one of $a + (b + xn)(x = 0, 1, \ldots, d - 1)$ modulo dn, so (iii) holds. Hence, there is exactly one of (i), (ii), (iii) holds. A', B' modulo dn satisfies the requirements of the question.

By (1) and (2), we know that all integers $n > 1$ except powers of 2 satisfy the conditions.

(3) When $n = 8$, it is easy to verify that $A = \{1, 2\}, B = \{3, 6\} A + B = \{4, 5, 7, 8\}$ satisfy the requirements. Combining with (2), all $2^k (k \geqslant 3)$ satisfy the requirements.

Since $A, B, A + B$ are non-empty sets, they modulo n have at least three distinct remainders, $n \geqslant 3$, and hence $n = 1, 2$ do not satisfy the conditions.

When $n = 4$, if A and B modulo 4 have only one remainder, then $A + B$ modulo 4 also has only one remainder and does not satisfy the requirements. If A or B modulo 4 has at least two different remainders, we might as well suppose $a_1, a_2 \in A$ are not congruent modulo 4. Taking any $b \in B$, then $a_1 + b$ is not congruent to $a_2 + b$ modulo 4, and thus $A + B$ modulo 4 has at least two remainders, which does not meet the requirements. Therefore, $n = 4$ also does not satisfy the conditions.

In summary, the required n is all positive integers except $1, 2$ and 4.

7 Let $n \geqslant 3$ be an integer. Given a convex n-gon P. An assignment to each vertex of P a color from red, yellow or blue is called 'good' if every interior point of P lies in the interior or on a triangle whose vertices are those of P with three different colors. Find the number of distinct good coloring methods.

Note: Two coloring methods are considered distinct if there exists a vertex assigned with different colors.

(Contributed by Xiao Liang)

Solution   The number of 'good' colorings is given by

$$f(n) = \begin{cases} 2^n - 2, & \text{if } n \text{ is odd,} \\ 2^n - 4, & \text{if } n \text{ is even.} \end{cases}$$

Let the vertices of P be labeled sequentially as $A_1, A_2, \ldots, A_n$, and the subscripts are modulo n. First of all, note that good coloring is exactly the coloring method in which all adjacent vertices are of different colors and all three colors are present.

Clearly, a good coloring uses all three colors. Next, we prove that in a good coloring, adjacent vertices have different colors. Otherwise, suppose adjacent vertices A_i, A_{i+1} are of the same color. Take a point X inside P that is very close to the midpoint of $A_i A_{i+1}$, and then X is only included in the interior of a triangle with A_i, A_{i+1} and some other point as vertices, but the three vertices of such a triangle are not of different colors. , so such coloring is not good.

In the following we will show that coloring where adjacent vertices are of different colors and all three colors are present is good. We use induction. When $n = 3$, the three vertices must have different colors, so this coloring method is good. Assume this conclusion holds for any less than n, and we will prove that this holds for convex n-gons. Take three vertices A, B, C with distinct colors (without loss of generality, assume $A = A_1, B = A_i$, and the colors of the three points are red, yellow, and blue, respectively). If the chosen points lie on the side or inside $\triangle ABC$, it satisfies the requirements, so it is only necessary to discuss the parts of P that corresponds to the exterior of each of the three diagonals AB, BC, CA separately (if one of the diagonals is a side of P, there is no need to discuss that side). In the following, we consider the convex polygon corresponding to the AB side, i.e., $A_1 A_2 \ldots A_i$, and the same applies to the remaining two cases.

If there is a blue point in $A_2 A_3 \cdots A_{i-1}$ that is a different color from A and B, the inductive hypothesis states that this point or any point on the boundary lies on the boundary or inside of some different-colored triangle.

If there are no blue points in $A_2 A_3 \cdots A_{i-1}$, since adjacent points are always of different colors, there must be that $A_2 A_4 \cdots A_i$ is blue (i is even), while $A_3 A_5 \cdots A_{i-1}$ is red. At this point $\triangle B A_1 A_2, \triangle B A_2 A_3, \ldots,$

$\triangle BA_{i-1}A_i$ are all different-colored triangles which cover the convex polygon $A_1A_2 \cdots A_i$. This completes the inductive proof.

Finally, we calculate the number of good coloring methods. For this we need not to consider the geometry shape. First we calculate the number of coloring methods in which adjacent points have different colors, denoted as D_n. It is clear that $D_2 = D_3 = 6$ (a digon with adjacent points of different colors is understood as two points of different colors). For $n \geqslant 4$, it is divided into two cases:

- A_n and A_2 are of different colors, and then the color of A_1 are uniquely determined by the colors of A_2 and A_n.
- A_n and A_2 are of the same color, and then we can combine A_n and A_2 into B. Then $BA_3 \cdots A_{n-1}$ is a $n - 2$ polygon with adjacent points of different colors.

Thus, we get

$$D_n = D_{n-1} + 2D_{n-2}.$$

Using the characteristic roots, we obtain the general formula $D_n = 2^n + 2 \cdot (-1)^n$.

Returning to the original problem, we need to subtract from the above calculations the colorings that use only two colors. This only occurs when n is even, and there are exactly 6 such colorings. Therefore, the number of good colorings is

$$f(n) = \begin{cases} D_n = 2^n - 2, & \text{if } n \text{ is odd}, \\ D_n - 6 = 2^n - 4, & \text{if } n \text{ is even}. \end{cases}$$

8   Let $x_1, x_2, \ldots, x_{11}$ be non-negative real numbers whose sum is equal to 1. For $i = 1, 2, \ldots, 11$, denote

$$y_i = \begin{cases} x_i + x_{i+1}, & \text{if } i \text{ is odd}, \\ x_i + x_{i+1} + x_{i+2}, & \text{if } i \text{ is even}, \end{cases}$$

where $x_{12} = x_1$. Let $F(x_1, x_2, \ldots, x_{11}) = y_1 y_2 \cdots y_{11}$.

Prove that when F achieves its maximum, there must be $x_6 < x_8$.

(Contributed by Qu Zhenhua)

Solution   Consider F as a function of $x_1, x_2, \ldots, x_{11}$ is a continuous function, it has a maximum on the bounded closed set

$$\Omega = \left\{ (x_1, x_2, \ldots, x_{11}) \in (\mathbf{R}_{\geqslant 0})^{11} \mid x_1 + x_2 + \cdots + x_{11} = 1 \right\}.$$

Suppose F takes the maximum at $(a_1, a_2, \ldots, a_{11}) \in \Omega$, and it is obvious that $F > 0$ at this time.

(1) $a_3 = a_5 = a_7 = a_9 = a_{11} = 0$.

If $a_3 > 0$, let $a_3' = 0, a_4' = a_3 + a_4$. For the other i, let $a_i' = a_i$, so the sum remains unchanged. We will prove that $F(a_1, a_2, \ldots, a_{11}) < F(a_1', a_2', \ldots, a_{11}')$, which is equivalent to

$$(a_2 + a_3 + a_4)(a_3 + a_4)(a_4 + a_5 + a_6)$$
$$< (a_2' + a_3' + a_4')(a_3' + a_4')(a_4' + a_5' + a_6').$$

By $a_3 + a_4 = a_3' + a_4'$ and $a_4 < a_4'$, if follows that the above inequality is established. This is inconsistent with F taking the maximum at $(a_1, a_2, \ldots, a_{11})$, so $a_3 = 0$. Similarly, we can prove $a_5 = a_7 = a_9 = a_{11} = 0$.

In the analytic expression of F, let $x_3 = x_5 = x_7 = x_9 = x_{11} = 0$. Now we consider F as a function of $x_1, x_2, x_4, \ldots, x_{10}$:

$$F = (x_1 + x_2)(x_2 + x_4)x_4(x_4 + x_6)x_6(x_6 + x_8)x_8(x_8 + x_{10})x_{10}(x_{10} + x_1)x_{10},$$

and then F takes its maximum at $(a_1, a_2, a_4, a_6, a_8, a_{10})$. It is obvious that $a_1, a_4, a_6, a_8, a_{10} > 0$.

(2) $a_2 = 0$.

We use proof by contradiction. Assume $a_2 > 0$. If $a_1 \leqslant a_4$, let $a_1' = a_1 + a_2, a_2' = 0$. For the other i, let $a_i' = a_i$, so the sum remains unchanged. We will prove

$$F(a_1, a_2, a_4, \ldots, a_{10}) < F(a_1', a_2', a_3', \ldots, a_{10}')$$
$$\Leftrightarrow a_1(a_{10} + a_1)(a_1 + a_2)(a_2 + a_4) < a_1'(a_{10}' + a_1')(a_1' + a_2')(a_2' + a_4').$$

Since $a_1 < a_1', a_{10} + a_1 < a_{10}' + a_1'$, we also only need to prove that

$$a_1(a_1 + a_2)(a_2 + a_4) \leqslant a_1'(a_1' + a_2')(a_2' + a_4') = (a_1 + a_2)(a_1 + a_2)a_4,$$
$$\Leftrightarrow a_1(a_2 + a_4) \leqslant a_4(a_1 + a_2) \Leftrightarrow a_1a_2 \leqslant a_4a_2$$

hold. If $a_1 \geqslant a_4$, let $a_4' = a_4 + a_2, a_2' = 0$. For the other i, let $a_i' = a_i$, and then similarly we have

$$F(a_1, a_2, a_4, \ldots, a_{10}) < F(a_1', a_2', a_4', \ldots, a_{10}').$$

Both are inconsistent with the fact that F takes its maximum at $(a_1, a_2, a_4, a_6, a_8, a_{10})$. Thus, $a_2 = 0$.

In the analytic expression of F, let $x_2 = 0$. Now we consider F as a function of $x_1, x_4, x_6, x_8, x_{10}$:

$$F = x_4^2(x_4 + x_6)x_6(x_6 + x_8)x_8(x_8 + x_{10})x_{10}(x_{10} + x_1)x_1^2,$$

and then F takes its maximum at $(a_1, a_4, a_6, a_8, a_{10})$.

(3) $a_1 = a_4, a_6 = a_{10}$.

Let $a_1' = a_4' = \frac{1}{2}(a_1 + a_4), a_6' = a_{10}' = \frac{1}{2}(a_6 + a_{10}), a_8' = a_8$, which keeps the sum unchanged. By the arithmetic-geometric mean inequality, we have

$$a_1^2 a_4^2 \leqslant a_1'^2 a_4'^2, \quad (a_4 + a_6)(a_{10} + a_1) \leqslant (a_4' + a_6')(a_{10}' + a_1'),$$

$$a_6 a_{10} \leqslant a_6' a_{10}', \quad (a_6 + a_8)(a_8 + a_{10}) \leqslant (a_6' + a_8')(a_8' + a_{10}').$$

And by $a_8 = a_8'$, multiplying these inequalities together gives

$$F(a_1, a_4, \dots, a_{10}) \leqslant F(a_1', a_4', \dots, a_{10}').$$

Since F takes its maximum at $(a_1, a_4, \dots, a_{10})$, the equality of the above equation is established, which implies $a_1 = a_4, a_6 = a_{10}$.

In the analytic expression of F, substitute x_4 with x_1 and x_{10} with x_6, so now we consider F as a function of x_1, x_6, x_8:

$$F = x_1^4 x_6^2 x_8 (x_1 + x_6)^2 (x_6 + x_8)^2,$$

and $2x_1 + 2x_6 + x_8 = 1$.

(4) Next we solve the system of equations (this system of equations is obtained by matching the coefficients of the AM-GM equality, which will be used in (5))

$$\frac{2}{u} + \frac{1}{u+v} = \frac{1}{v} + \frac{1}{u+v} + \frac{1}{v+1} = 1 + \frac{2}{v+1}, \quad u, v > 0.$$

We prove that there are unique positive real solutions u, v with $v < 1$.

From the first equation, we get $\frac{2}{u} = \frac{1}{v} + \frac{1}{v+1}$, and thus $u = \frac{2v(v+1)}{2v+1}$. And from the second equation, we get $\frac{1}{v} + \frac{1}{u+v} = 1 + \frac{1}{v+1}$, Substituting u and simplifying in terms of v, we obtain

$$4v^3 + 5v^2 - 4v - 4 = 0.$$

Let $f(t) = 4t^3 + 5t^2 - 4t - 4$. Then $f'(t) = 12t^2 + 10t - 4$ is negative first and then positive on $[0, +\infty)$, and hence $f(t)$ first increases and then decreases on $[0, +\infty)$. Since $f(0) = -4 < 0, f(1) = 1 > 0$, f has a unique solution $v \in (0, 1)$ on $[0, +\infty)$, and hence u is uniquely determined.

(5) Let $u, v > 0$ be the solution that satisfies the system of equations in (4), and let $\frac{2}{u} + \frac{1}{u+v} = k$. Using the arithmetic-geometric mean inequality, we have

$$F(x_1, x_6, x_8)$$

$$= \left(\frac{x_1}{u}\right)^4 \left(\frac{x_6}{v}\right)^2 x_8 \left(\frac{x_1 + x_6}{u+v}\right)^2 \left(\frac{x_6 + x_8}{v+1}\right)^2 u^4 v^2 (u+v)^2 (v+1)^2$$

$$\leq \frac{u^4 v^2 (u+v)^2 (v+1)^2}{11^{11}}$$

$$\times \left(\frac{4}{u} x_1 + \frac{2}{v} x_6 + x_8 + \frac{2}{u+v}(x_1 + x_6) + n \frac{2}{v+1}(x_6 + x_8)\right)^{11}$$

$$= \frac{u^4 v^2 (u+v)^2 (v+1)^2}{11^{11}}$$

$$\times \left(\left(\frac{4}{u} + \frac{2}{u+v}\right) x_1 + \left(\frac{2}{v} + \frac{2}{u+v} + \frac{2}{v+1}\right) x_6 \, n + \left(1 + \frac{2}{v+1}\right) x_8\right)^{11}$$

$$= \frac{u^4 v^2 (u+v)^2 (v+1)^2}{11^{11}}(2kx_1 + 2kx_6 + kx_8)^{11}$$

$$= \frac{u^4 v^2 (u+v)^2 (v+1)^2 k^{11}}{11^{11}}.$$

The above equality holds if and only if $x_1 : x_6 : x_8 = u : v : 1$, that is, $x_8 = \frac{1}{2u+2v+1}, x_6 = \frac{v}{2u+2v+1}, x_1 = \frac{u}{2u+2v+1}$. Hence, F takes its maximum if and only if $x_2 = x_3 = x_5 = x_7 = x_9 = x_{11} = 0, x_1 = x_4 = \frac{u}{2u+2v+1}, x_6 = x_{10} = \frac{v}{2u+2v+1}, x_8 = \frac{1}{2u+2v+1}$. At this point, $x_6 = vx_8 < x_8$.

5

The 19th China Southeastern Mathematical Olympiad

The 19th "Wenshan Cup" China Southeastern Mathematical Olympiad (CSMO) was held from 31st July to 4th August 2022, at Bailuzhou Middle School in Ji'an city. A total of 1692 students from 151 high schools across the country participated in this event. After intense competition, Jiangxi University of Technology High School and Shanghai Huayu Private Middle School won the first place in the total team score of the 10th Grade and 11th Grade, respectively, and a total of 161 students won gold medals. The main examination committee included: Tao Pingsheng,[1] Zhang Pengcheng, Li Shenghong, Yang Xiaoming, Xiong Bin, Wu Quanshui, Lin Tianqi, Gong Fuzhou, Hu Zhiming, Wang Jiajun, Liu Bin, Dong Qiuxian, Wu Genxiu, He Yijie, Luo Ye, and Song Yuanlong.

This was a vibrant and rewarding event—the organizing committee prepared a variety of academic and cultural activities for the participants. Through academic lectures, interviews with famous experts, highlights of the live broadcast of activities, wonderful cultural performances, as well as a visit to the Millennium Academy and Millennium Kiln, displaying the traditional and modern convergence and integration of Luzhong. It narrated the stories of the flourishing development and rapid changes in Jinggangshan, opening a window for participants from various regions to learn about Ji'an and Luhong.

---

[1] Tao Pingsheng, January 1946–10th March 2023.

Bailuzhou Middle School of Ji'an City is renowned at home and abroad for its long history of educational operation, elegant study environment and remarkable educational benefits. The school has a long history of literary style, and its predecessor is the Bailuzhou Academy, one of the four major academies in Jiangxi founded by Mr. Jiang Wanli, a famous patriotic prime minister and educator, in 1241, which is the alma mater of Wen Tianxiang. The school upholds the spirit of 'inheriting civilization and pursuing excellence', and has nurtured generation after generation of talented individuals who have made practical contributions to society. These include prominent figures in the scientific community, outstanding educators, business elites, and accomplished individuals in arts and sports.

10th Grade
First Day
(8:00–12:00; 2nd August 2022)

1   Suppose positive sequence $\{a_n\}$ satisfies $a_1 = 1 + \sqrt{2}$ and $(a_n - a_{n-1})(a_n + a_{n-1} - 2\sqrt{n}) = 2(n \geqslant 2)$.

(1) Find the general term formula for sequence $\{a_n\}$;

(2) find the set of all positive integers n such that $\lfloor a_n \rfloor = 2022$, where $\lfloor x \rfloor$ denotes the greatest integer less than or equal to x.

(Contributed by Wu Genxiu)

Solution

(1) By $(a_2 - 1 - \sqrt{2})(a_2 + 1 + \sqrt{2} - 2\sqrt{2}) = 2$ and $a_n > 0$, we have

$$a_2 = \sqrt{2} + \sqrt{3}.$$

Assuming $a_n = \sqrt{n} + \sqrt{n+1}$, we prove this for $n + 1$ as follows:

By $(a_{n+1} - a_n)(a_{n+1} + a_n - 2\sqrt{n+1}) = 2$, substituting $a_n = \sqrt{n} + \sqrt{n+1}$ into this we have

$$a_{n+1}^2 - 2\sqrt{n+1}\,a_{n+1} - 1 = 0.$$

And according to $a_{n+1} > 0$, the solution is $a_{n+1} = \sqrt{n+1} + \sqrt{n+2}$.

(2) Note that $a_n^2 = (\sqrt{n} + \sqrt{n+1})^2 = 2n + 1 + 2\sqrt{n(n+1)}$. By

$$n < \sqrt{n(n+1)} < n + \frac{1}{2},$$

it follows that $\sqrt{4n+1} < \sqrt{n} + \sqrt{n+1} < \sqrt{4n+2}$.

If $[\sqrt{n} + \sqrt{n+1}] \neq [\sqrt{4n+2}]$, then there exists $k \in \mathbf{N}_+$ such that

$$\sqrt{4n+1} < k \leqslant \sqrt{4n+2},$$

namely, $4n + 1 < (\sqrt{n} + \sqrt{n+1})^2 < k^2 \leqslant 4n + 2$. Hence, $4n + 2 = k^2$ contradicts square number $\equiv 0$ or $1 \pmod 4$. Therefore,

$$\lfloor a_n \rfloor = \lfloor \sqrt{n} + \sqrt{n+1} \rfloor = \lfloor \sqrt{4n+2} \rfloor.$$

From all positive integers n that satisfy $\lfloor a_n \rfloor = 2022$, there must be $\lfloor \sqrt{4n+2} \rfloor = 2022$, that is,

$$2022 \leqslant \sqrt{4n+2} < 2023,$$

which implies

$$1022120.5 \leqslant n < 1023131.75.$$

The set of all positive integers n that satisfy $\lfloor a_n \rfloor = 2022$ is

$$\{n \mid 1022121 \leqslant n \leqslant 1023131, n \in \mathbf{N}_+\},$$

which has a total of 1011 elements.

2. As shown in Fig. 5.1, in acute-angled triangle ABC, $AB > AC$. H is the orthocenter, and AM is the midline. $BE \perp AC$ at point E, $CF \perp AB$ at point F. Point D lies on side BC, satisfying $\angle CAD = \angle BAM$ and $\angle ADH = \angle MAH$. Prove that EF bisects segment AD.

(Contributed by Zhang Pengcheng)

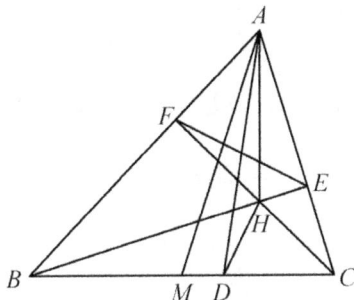

Fig. 5.1

Solution As shown in Fig. 5.2, denote the circumcircles of $\triangle ABC$ and $\triangle HBC$ as circle O and circle P, respectively. By the properties of the orthocenter, we know that circle O and circle P are symmetric about point M.

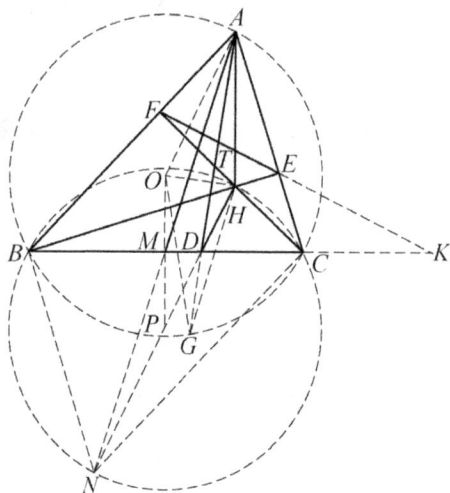

Fig. 5.2

Assume that the extension of AM intersects circle P at point N, and then $AM = MN$. Connect HN, BN, CN.

Thus, quadrilateral $ABNC$ is a parallelogram. Hence, $BN \perp BE$, and point P lies on HN.

And since $OP \parallel AH$ and $OP = AH$, $HP \parallel AO$, i.e., $HN \parallel AO$.

It is easy to see that $\angle BAO = \angle CAH$. And since $\angle BAM = \angle CAD$, $\angle OAM = \angle HAD$.

Therefore, $\angle ADH = \angle MAH = \angle DAO$, and hence $HD \parallel OA$, that is, point D lies on HN.

Suppose the extension of AD intersects circle O at point G. Then

$$AD \cdot DG = BD \cdot DC = HD \cdot DN,$$

and thus points A, N, G, H are concyclic.

Therefore, $\angle AGH = \angle ANH = \angle OAM = \angle GAH$. And since $OA = OG$,

$$AD \perp OH.$$

Suppose AD and EF intersect at point T, and lines EF and BC intersect at point K. Given that points B, C, E, F are concyclic, it follows that

$$\angle BAO = 90° - \angle ACB = 90° - \angle AFE,$$

and hence $AO \perp EF$.

And since $AD \perp OH, AH \perp BC, \angle AOH = \angle KTD, \angle AHO = \angle KDT$, so $\triangle AOH \backsim \triangle KTD$.

Hence,

$$\frac{KT}{KD} = \frac{AO}{AH} = \frac{1}{2\cos A} = \frac{AB}{2AE}. \qquad \text{①}$$

In $\triangle KCE$, by the sine rule, it follows that

$$\frac{KC}{KE} = \frac{\sin \angle KEC}{\sin \angle KCE} = \frac{\sin \angle ABC}{\sin \angle ACB} = \frac{AC}{AB}. \qquad \text{②}$$

Line TEK intercepts $\triangle ADC$. According to Menelaus' theorem, we have

$$\frac{AT}{TD} \cdot \frac{DK}{KC} \cdot \frac{CE}{EA} = 1. \qquad \text{③}$$

Line DCK intercepts $\triangle ATE$. According to Menelaus' theorem, there is

$$\frac{TD}{DA} \cdot \frac{AC}{CE} \cdot \frac{EK}{KT} = 1. \qquad \text{④}$$

Multiplying ③ and ④, we get

$$\frac{AT}{DA} \cdot \frac{DK}{KC} \cdot \frac{AC}{EA} \cdot \frac{EK}{KT} = 1,$$

that is,

$$\frac{AT}{AD} = \frac{KT}{KD} \cdot \frac{KC}{KE} \cdot \frac{AE}{AC}. \qquad \text{⑤}$$

Substituting ① and ② into ⑤, we have $\frac{AT}{AD} = \frac{AB}{2AE} \cdot \frac{AC}{AB} \cdot \frac{AE}{AC} = \frac{1}{2}$. Therefore, EF bisects segment AD.

3 Let x_i be an integer greater than 1. Denote $f(x_i)$ as the greatest prime factor of x_i. Let $x_{i+1} = x_i - f(x_i)$ (i is a natural number).

(1) Prove that for any integer x_0 greater than 1, there exists a natural number $k(x_0)$ such that $x_{k(x_0)+1} = 0$;

(2) Let $V(x_0)$ denote the number of distinct numbers in $f(x_0)$, $f(x_1), \ldots, f(x_{k(x_0)})$. Find the greatest number in $V(2), V(3), \ldots$, $V(781)$ and give reasons. (*Note*: Bailuzhou Academy was founded in 1241 AD, and has a history of 781 years.)

(Contributed by Luo Ye)

Solution　Let $p_i = f(x_i)$ be a prime number.

(1) By definition, for any i, if $x_i \in \mathbf{N}_+, x_i > 1$, then it is easy to see that $x_i > x_{i+1} \geqslant 0$ is a descending integer sequence and $p_i \mid x_{i+1}$. Therefore, for any initial $x_0 \in \mathbf{N}_+, x_0 > 1$, this descent can only be performed a finite number of times. So there must be some natural number k such that $x_k \geqslant 2, x_{k+1} < 2$. But since $x_{k+1} \geqslant 0$ and $p_k \mid x_{k+1}, x_{k+1}$ cannot be 1, there must be $x_{k+1} = 0$.

(2) It is easy to know that p_i is an increasing sequence while x_i is a strictly decreasing sequence, $i = 0, 1, \ldots, k$. Let $i_1, \ldots, i_s$ be the number in $1, \ldots, k$ such that $p_{i_l} > p_{i_l - 1}, l = 1, 2, \ldots, s$. Thus, $V(x) = s + 1$. Denote $q_0 = p_0, q_l = p_{i_l}, l = 1, 2, \ldots, s$. We know that $q_0, \ldots, q_s$ are all different prime numbers in $f(x_0), \ldots, f(x_k)$, and $q_s > q_{s-1} > \cdots > q_0$. It is easy to know that $q_l q_{l-1} \mid x_{i_l}, l = 1, 2, \ldots, s$. Let $k_l = \frac{x_{i_l}}{q_l q_{l-1}}$. It can be seen that $k_s < k_{s-1} < \cdots < k_1$ is a strictly decreasing sequence of positive integers.

First, we prove $s \leqslant 4$. Now we assume $s \geqslant 5$ inversely, namely, $V(x_0) \geqslant 6$.

Therefore, there is $q_5 \geqslant 13, q_4 \geqslant 11$, and then $x_0 > 13 \times 11 = 143$.

If $q_0 = 2$, then x_0 is a power of 2. Since $x_0 = 781, x_0 \in \{256, 512\}$. But it can be verified that $V(256) = 2, V(512) = 3$ is inconsistent with $V(x_0) \geqslant 6$.

Similarly, $q_0 \neq 3$ can be verified.

Thus, $q_0 \geqslant 5$, and hence $q_5 \geqslant 19, q_4 \geqslant 17, \ q_3 \geqslant 13, q_2 \geqslant 11, q_1 \geqslant 7$. So we have

$$x_0 > 17 \times 19 = 323.$$

Therefore, $k_5 \leqslant 2, k_4 \leqslant 3$.

We can confirm that $q_{s-1} \leqslant 23$; otherwise, we would have $x_0 > q_{s-1} q_s \geqslant 29 \times 31 > 781$, a contradiction.

By definition, we know that $x_{i_l}, x_{i_l + 1}, \ldots, x_{i_{l+1}}$ is an arithmetic sequence and covers all multiples of q_l from $q_l(q_{l+1} k_{l+1})$ to $q_l(q_{l-1} k_l)$. Thus, we have the following properties:

Property (i):　$q_{l+1} k_{l+1} < q_{l-1} k_l$ and the maximum prime factor of every number in $q_{l+1} k_{l+1} + 1, \ldots, q_{l-1} k_l$ is not greater than q_l.

Property (ii):　There is at least one prime number for every 11 consecutive numbers between 1 and 118.

Property (iii):　If $q_{l-1} \geqslant 11$ and $q_{l-1} k_l \leqslant 118$, then $q_{l-1} k_l$ is exactly the first number greater than or equal to $q_{l+1} k_{l+1} + 1$ and divisible by q_{l-1}. Specifically, define $I_{l+1} = \{q_{l+1} k_{l+1} + 1, q_{l+1} k_{l+1} + 2, \ldots, q_{l-1} k_l\}, l = 1, 2, \ldots, s - 1$.

Then by properties (i) and (ii), we know that none of the numbers in I_{l+1} can be prime numbers, nor can they have prime factors larger than q_l.

We can prove $q_3 = 13$.

Since $q_3 \geqslant 13$, all prime factors of $q_5 k_5 + 1, \ldots, q_3 k_4$ are not greater than q_4.

If $q_3 \geqslant 17$, then $q_4 \geqslant 19, q_5 \geqslant 23$, so k_5 must be 1. Therefore,

$$k_5 q_4 q_5 \leqslant x_0 - q_0 - q_1 - q_2 - q_3 - q_4 < 760,$$

so $q_5 \leqslant 40$. Thus, $q_5 \in \{23, 29, 31, 37\}$.

(a) If $q_5 = 23$, since $q_5 k_5 + 1 = 24 < 34 \leqslant q_3 k_4$, $29 \in I_5$ while $29 > q_5 > q_4$, which contradicts property (i).

(b) If $q_5 = 29$, then $q_5 k_5 + 1 = 30$ cannot be divisible by q_3, while $q_5 k_5 + 1 = 31 \in I_5$, which contradicts property (iii).

(c) If $q_5 = 31$, then $q_5 k_5 + 1 = 32$. If $k_4 = 3$, then there is $37 \in I_5$, which contradicts property (iii).

Similarly, consider $q_4 k_4 + 1 = 39$. If $41 \in I_4$, then it is inconsistent with property (iii), so $q_2 = 13, k_3 = 3$.

Consider the sequence starting with $q_3 k_3 + 1 = 52$. The next number divisible by $q_1 \in \{7, 11\}$ is 55. However, $53 \in [51, 55]$ is a prime number, and $53 > q_2$, a contradiction.

(d) If $q_5 = 37$, then $q_5 k_5 + 1 = 38$. If $q_3 > 19$, then $41 \in I_5$, a contradiction. Therefore, $q_3 = 19, q_4 \geqslant 23$, so $x_0 \geqslant q_4 q_5 > 781$, a contradiction.

Thus, there must be $q_3 = 13$. Consequently, $q_2 = 11, q_1 = 7, q_0 = 5$. Hence, $k_3 < 781/(11 \times 13) < 6$, and thus $3 \leqslant k_3 \leqslant 5$. If $k_3 = 3$, then $q_3 k_3 + 1 = 39$. If $41 \in I_3$, then $41 > q_2$, a contradiction. If $k_3 = 4$, then $q_3 k_3 + 1 = 53$ is a prime number, contradicting property (iii). If $k_3 = 5$, then $x_{i_3} = 5 \times 11 \times 13$, and it is easy to see that $V(x_{i_3}) = 2$, so $V(x_0) = V(x_{i_3}) + 3 = 5 < 6$, a contradiction.

In summary, $V(x_0) \leqslant 5$ and greatest number can be achieved when $x_0 = 320$. Therefore, the greatest number is 5.

4 Given integers $m, n \geqslant 2$. Color each cell of a grid table S of m rows and n columns either red or blue such that the following conditions hold: for any two cells in the same row, if they are both colored red, then the two columns they belong to are such that all cells in one column are colored red, and all cells in the other column are colored blue. Find the number of different coloring methods.

(Contributed by He Yijie)

Solution   Let (i, j) denote the intersection cell of row i and column j, where $1 \leqslant i \leqslant m, 1 \leqslant j \leqslant n$.

The analysis is divided into the following three cases:

Case 1: There is no column in S where all cells are red.

In this case, by the given conditions, each row in S has at most one red cell. Therefore, there are $n + 1$ coloring methods for each row (1 way for all blue cells, n ways for having at least one red cell). By the multiplication principle, a total of $(n + 1)^m$ coloring methods are obtained. Excluding the cases where all red cells are concentrated in one column (n cases), there are $(n + 1)^m - n$ coloring methods that meet the requirements.

Case 2: There is exactly one column in S where all cells are red.

Assume that column k is all red cells, where $k \in \{1, 2, \ldots, n\}$. Consider the grid table $S(\hat{k})$ with m rows and $n - 1$ columns composed of the remaining grids in S, and then there is no column in $S(\hat{k})$ where all cells are red. Similar to case 1, it can be seen that each row of $S(\hat{k})$ has at most one red cell, and $S(\hat{k})$ has $n^m - (n - 1)$ coloring methods.

For each coloring method, the coloring methods of S obtained after adding the red cells in the column k satisfy the conditions. In fact, if there are two red squares P and Q in the same row, since $S(\hat{k})$ has at most one red cell in that row, one of P and Q must be in column k and the other one in $S(\hat{k})$. At this point, the two rows to which P and Q belong indeed have one column (column k) that all cells are red, meeting the requirements.

Note that there are n ways to choose k, by the addition principle, there are $n \cdot (n^m - n + 1)$ coloring methods that meet the requirements.

Case 3: There are two columns in S where all cells are red.

Without loss of generality, assume column j and column k are entirely red. Then, the red cells $(1, j), (1, k)$ clearly do not conform to the problem's requirements. Therefore, such a coloring method does not exist.

Combining Cases 1, 2, and 3, the number of coloring methods that meet the requirements is

$$(n + 1)^m - n + n \cdot (n^m - n + 1) = (n + 1)^m + n^{m+1} - n^2.$$

10th Grade
Second Day
(8:00–12:00; 3rd August 2022)

5 Suppose positive sequences $\{a_n\}, \{b_n\}$ satisfy: $a_1 = b_1 = 1$, $b_n = a_n b_{n-1} - \frac{1}{4} (n \geq 2)$.

Find the minimum value of $4\sqrt{b_1 b_2 \cdots b_m} + \sum_{k=1}^{m} \frac{1}{a_1 a_2 \cdots a_k}$, where m is a given positive integer.

(Contributed by Li Shenghong)

Solution By the assumption, there is $a_n = \frac{b_n + \frac{1}{4}}{b_{n-1}}, n \geq 2$.

Thus, $a_1 a_2 \cdots a_n = a_2 \cdots a_n = \frac{\left(b_2 + \frac{1}{4}\right)\left(b_3 + \frac{1}{4}\right) \cdots \left(b_n + \frac{1}{4}\right)}{b_1 b_2 \cdots b_{n-1}}$, that is,

$$\frac{b_n}{a_1 a_2 \cdots a_n} = \frac{b_n}{a_2 \cdots a_n} = \frac{b_2 \cdots b_n}{\left(b_2 + \frac{1}{4}\right)\left(b_3 + \frac{1}{4}\right) \cdots \left(b_n + \frac{1}{4}\right)}$$

$$= \frac{1}{\left(1 + \frac{1}{4b_2}\right)\left(1 + \frac{1}{4b_3}\right) \cdots \left(1 + \frac{1}{4b_n}\right)} \leqslant \sqrt{b_1 b_2 \cdots b_n}.$$

And by

$$\sum_{k=1}^{m} \frac{1}{a_1 a_2 \cdots a_k} = \frac{1}{a_1} + \sum_{k=2}^{m} \frac{1}{a_1 a_2 \cdots a_k} = 1 + \sum_{k=2}^{m} \frac{1}{a_1 a_2 \cdots a_k} 4(a_k b_{k-1} - b_k)$$

$$= 1 + 4\sum_{k=2}^{m} \left(\frac{b_{k-1}}{a_1 a_2 \cdots a_{k-1}} - \frac{b_k}{a_1 a_2 \cdots a_k} \right)$$

$$= 1 + 4\left(\frac{b_1}{a_1} - \frac{b_m}{a_1 a_2 \cdots a_m} \right)$$

$$= 5 - \frac{4b_m}{a_1 a_2 \cdots a_m} \geqslant 5 - 4\sqrt{b_1 b_2 \cdots b_m},$$

it follows that $4\sqrt{b_1 b_2 \cdots b_m} + \sum_{k=1}^{m} \frac{1}{a_1 a_2 \cdots a_k} \geqslant 5$.

Let $b_2 = b_3 = \cdots = b_m = \frac{1}{4}$. By $a_n = \frac{b_n + \frac{1}{4}}{b_{n-1}}$, we can get $a_2 = \frac{1}{2}$, $a_3 = a_4 = \cdots = a_m = 2$, and hence

$$4\sqrt{b_1 b_2 \cdots b_m} + \sum_{k=1}^{m} \frac{1}{a_1 a_2 \cdots a_k}$$

$$= 4\sqrt{\left(\frac{1}{4}\right)^{m-1}} + 1 + 2 + 1 + \frac{1}{2} + \cdots + \left(\frac{1}{2}\right)^{m-3}$$

$$= 5.$$

Therefore, the minimum value of $4\sqrt{b_1 b_2 \cdots b_m} + \sum_{k=1}^{m} \frac{1}{a_1 a_2 \cdots a_k}$ is 5.

6 As shown in Fig. 5.3, in $\triangle ABC$, H is the orthocenter. With H as the center, the circle passing through point A intersects sides AC and AB at points D and E, respectively, where D and E are different from A. The orthocenter of $\triangle ADE$ is H', and the extension of AH' intersects DE at point F. Point P lies inside quadrilateral $BCDE$, satisfying $\triangle PDE \backsim \triangle PBC$ (the vertices are correspondingly ordered). Suppose lines HH' and PF intersect at point K. Prove that points A, H, P, K are concyclic.

(Contributed by Lin Tianqi)

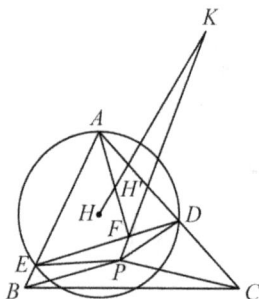

Fig. 5.3

Solution As shown in Fig. 5.4, suppose the extension of AF intersects $\odot H$ at point A'. Construct the circumcircle ω of $\triangle AHA'$. Suppose line HH' and ω intersect at another point K' other than H. We will prove that points A, H, P, A', K are concyclic. This is proved in three steps: (1) Point P lies on ω; (2) Point K' lies on the circumcircle of $\triangle PDE$; (3) K' coincides with K.

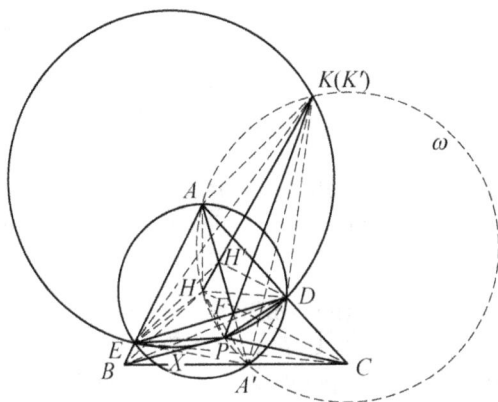

Fig. 5.4

(1) Prove that point P lies on ω.

By the given condition $\triangle PDE \backsim \triangle PBC$, there is

$$\angle A'DE = \angle EAF = 90° - \angle AED = \angle HAD = \angle HBC,$$

namely, $\angle A'DE = \angle HBC$. Similarly, there is $\angle A'ED = \angle HCB$, and thus $\triangle A'DE \backsim \triangle HBC$. Therefore,

$$\angle HPA' = \angle(BC, DE) = \angle HAA'.$$

Thus, points A, H, P, A' are concyclic, and hence point P lies on ω.

(2) Prove that point K' lies on the circumcircle of $\triangle PDE$.

By $HA = HA'$ and inscribed angle theorem, we have

$$\angle HAH' = \angle AA'H = \angle AK'H,$$

and thus $\triangle HAH' \backsim \triangle HK'A$. Therefore, $HA^2 = HH' \cdot HK'$. And since $HD = HE = HA$,

$$HD^2 = HE^2 = HH' \cdot HK'.$$

Hence, $\triangle HDH' \backsim \triangle HK'D$, $\triangle HEH' \backsim \triangle HK'E$, and thus

$$\angle DK'E = \angle HK'D + \angle HK'E = \angle HDH' + \angle HEH'$$

$$= \angle DHE - \angle DH'E = 2\angle BAC - (180° - \angle BAC)$$

$$= 3\angle BAC - 180°.$$

Suppose BD and CE intersect at point X. By the properties of Miquel points, we know that the circumcircles of $\triangle XDE$ and $\triangle XBC$ intersect at another point P other than X. Note that $AB = BD, AC = CE$, so

$$\angle DPE = \angle DXE = 360° - \angle DAE - \angle AEX - \angle ADX$$

$$= 360° - 3\angle BAC.$$

Therefore, $\angle DK'E + \angle DPE = 180°$, and then points D, P, E, K' are concyclic, which implies that point K' lies on the circumcircle of $\triangle PDE$.

(3) Prove that K' coincides with K.

Considering that PK' is the radical axis of circle ω and the circumcircle of $\triangle DPE$, and given that $\overline{FA} \cdot \overline{FA'} = \overline{FD} \cdot \overline{FE}$, point F is equidistant from both circles. Therefore, points P, F, K' are collinear.

Consequently, K' coincides with K.

Therefore, points A, H, P, K all lie on circle ω, which completes the proof.

⑦ Suppose a, b are positive integers. Prove that there are no positive integers on the interval $\left[\frac{b^2}{a^2+ab}, \frac{b^2}{a^2+ab-1}\right)$.

(Contributed by Yang Xiaoming)

Solution If $a \geqslant b$, then $\frac{b^2}{a^2+ab} < \frac{b^2}{a^2+ab-1} \leqslant \frac{b^2}{ab} \leqslant 1$, so $\left[\frac{b^2}{a^2+ab}, \frac{b^2}{a^2+ab-1}\right) \subset [0, 1)$. The proposition is true.

If $a < b$,

(1) When $a = 1$, the original interval is $\left[\frac{b^2}{1+b}, \frac{b^2}{b}\right) = \left[\frac{1}{1+b} + b - 1, b\right)$. Because b is a positive integer, there are no other integers between two consecutive integers $b - 1, b$.

(2) When $a > 1$, assume there exists a positive integer $b > a > 1$ and the interval $\left[\frac{b^2}{a^2+ab}, \frac{b^2}{a^2+ab-1}\right)$ contains positive integers. Take such (a, b) to minimize $a + b$.

Let $b = aq + r, q, r \in \mathbf{N}^*, 0 \leqslant r \leqslant a - 1$. Then

$$\frac{b^2}{a^2+ab} - (q-1) = \frac{(aq+r)^2}{a^2+a(aq+r)} - q + 1 = \frac{a^2 + (q+1)ar + r^2}{(q+1)a^2 + ar} > 0,$$

$$q + 1 - \frac{b^2}{a^2+ab-1} = q + 1 - \frac{(aq+r)^2}{a^2+a(aq+r)-1}$$

$$= \frac{(2q+1)a^2 - (q-1)ar - q - 1 - r^2}{(q+1)a^2 + ar - 1}$$

$$\geqslant \frac{(2q+1)a^2 - (q-1)a(a-1) - q - 1 - (a-1)^2}{(q+1)a^2 + ar - 1}$$

$$= \frac{(q+1)(a^2+a) - 2 - q}{(q+1)a^2 + ar - 1} > 0.$$

Therefore, $\left[\frac{b^2}{a^2+ab}, \frac{b^2}{a^2+ab-1}\right)$ is strictly contained in $(q-1, q+1)$. So $q \in \left[\frac{b^2}{a^2+ab}, \frac{b^2}{a^2+ab-1}\right)$ is equivalent to

$$\frac{(aq+r)^2}{a^2 + a(aq+r)} \leqslant q < \frac{(aq+r)^2}{a^2 + a(aq+r) - 1}. \qquad \text{①}$$

The left-hand side of equation ① is equivalent to $a^2q^2 + 2qra + r^2 \leqslant (q^2+q)a^2 + qra \Leftrightarrow \frac{r^2}{a^2-ar} \leqslant q$;

The right-hand side of equation ① is equivalent to $(q^2+q)a^2 + qra - q < a^2q^2 + 2qra + r^2 \Leftrightarrow q < \frac{r^2}{a^2-ar-1}$.

In summary, we can get positive integer $q \in \left[\frac{r^2}{a^2-ar}, \frac{r^2}{a^2-ar-1}\right)$.

It is obvious that $r \neq 0$, otherwise q would not exist.

Suppose $p = a - r$. Then $a = p + r, p, r \in \mathbf{N}_+$, and thus

$$q \in \left[\frac{r^2}{(p+r)^2 - (p+r)r}, \frac{r^2}{(p+r)^2 - (p+r)r - 1}\right),$$

namely, $q \in \left[\frac{r^2}{p^2+pr}, \frac{r^2}{p^2+pr-1}\right)$.

By definition, it is obvious that $p + r = a < a + b$, so (p, r) is also a set of positive integers that meet the requirements. However, $p + r < a + b$, and this is inconsistent with the minimality of $a + b$.

Therefore, there are no positive integers on the interval $\left[\frac{b^2}{a^2+ab}, \frac{b^2}{a^2+ab-1}\right)$.

8 Student John plays the following game: take a constant v greater than 1; for positive integer m, the interval between the mth round and the $m+1$th round is 2^{-m} units of time; in the mth round, a safety disk area with a radius of 2^{-m+1} is taken in the plane (including the boundary, the time to take the disk is negligible); after being taken, the safety disk area will keep the center of the disk unchanged for the rest of the game, with the radius decreasing at a constant rate v until it reaches zero, at which point the safety disk area is removed. If John can completely locate the safety disk area within the existing safety area in a round before the 100th round (including the 100th round), find the minimum value of $\left\lfloor \frac{1}{v-1} \right\rfloor$, where $\lfloor x \rfloor$ denotes the greatest integer that does not exceed x.

(Contributed by Luo Ye)

Solution

Lemma *Suppose the boundary of disk O with radius R is inscribed in convex quadrilateral $ABCD$, and let the lengths of AB, BC, CD and DA be a, b, c and d, respectively. Then the necessary and sufficient conditions for $R \geqslant 1$ are $x^4 - (2a^2 + 2b^2 - a^2b^2)x^2 + (a^2 - b^2)^2 \geqslant 0$, where x is the length of diagonal AC.*

Proof of the lemma   Let $\angle ABC = \theta$, and then $\angle ADC = \pi - \theta$. Then the length x of diagonal AC satisfies $x^2 = a^2 + b^2 - 2ab \cos\theta$ and $x^2 = c^2 + d^2 - 2cd \cos(\pi - \theta)$. From these two equations we have

$$x^2 = \frac{cd(a^2 + b^2) + ab(c^2 + d^2)}{ab + cd}.$$

Because this disk is also the circumcircle of $\triangle ABC$, by Heron's formula we have

$$R = \frac{abx}{4S_{\triangle ABC}} = \frac{abx}{\sqrt{(a + b + x)(a + b - x)(a + x - b)(b + x - a)}}.$$

From this, we conclude that the necessary and sufficient conditions for $R \geqslant 1$ are

$$a^2 b^2 x^2 \geqslant ((a + b)^2 - x^2)(x^2 - (a - b)^2),$$

namely,

$$x^4 - (2a^2 + 2b^2 - a^2b^2)x^2 + (a^2 - b^2)^2 \geqslant 0.$$

Denote $v = 1 + v_0, k = \left\lfloor \frac{1}{v-1} \right\rfloor \in \mathbf{N}_+$. We will prove that the minimum value of k is 18.

Denote the disk drawn by John at the beginning of the mth round as C_m. Therefore, at the beginning of the $m + 1$th round, if its radius has not yet become zero, the radius will have changed to $r_{m,1} = 2^{-m+1} - v \cdot \sum_{i=0}^{l-1} 2^{-m-i} = 2^{-m+1}\left(1 - v\left(1 - \frac{1}{2^l}\right)\right) = \frac{2^l - v(2^l - 1)}{2^{m+l-1}}$. Hence, at the beginning of the $m + l$th round, its radius becomes

$$r_{m,l} = \max\left\{0, \frac{2^l - v(2^l - 1)}{2^{m+l-1}}\right\} = \max\left\{0, \frac{1 - v_0(2^l - 1)}{2^{m+l-1}}\right\}.$$

It can be seen from this that when $v_0 \in \left[\frac{1}{2^{l_0+1}-1}, \frac{1}{2^{l_0}-1}\right)$, where $l_0 \in \mathbf{N}_+$, the number of disks on the plane at the start of any mth round is at most l_0. The radii of these disks are

$$\frac{1 - v_0(2^l - 1)}{2^{m-1}}, \quad l = 1, 2, \ldots, l_0.$$

The problem becomes a plane covering problem: whether it is possible to cover a disk with radius $\frac{1}{2^{m-1}}$ with l_0 disks with radius $\frac{1-v_0(2^l-1)}{2^{m-1}}, l = 1, 2, \ldots, l_0$. That is, can a disk of radius 1 be covered with l_0 disks with radius $v_0(2^l - 1), l = 1, 2, \ldots, l_0$.

Since if for $v > 1$, John has a winning strategy, then for any $v' \in (1, v)$, John also has a winning strategy. Therefore, for any k, if there exists $v_0 \in \left(\frac{1}{1+k}, \frac{1}{k}\right]$ such that John has a winning strategy, then for $k' > k$, there also exists $v_0' \in \left(\frac{1}{1+k'}, \frac{1}{k'}\right]$ such that John also has a winning strategy.

(1) We first prove that $k = 18$ is achievable. That is, prove that there exists $v_0 \in \left(\frac{1}{19}, \frac{1}{18}\right]$ such that John has a winning strategy. When $v_0 \in \left(\frac{1}{19}, \frac{1}{18}\right]$, $l_0 = 4$, so there are at most 4 disks left on the plane at any time, and their radii are $1 - v_0, 1 - 3v_0, 1 - 7v_0, 1 - 15v_0$, respectively.

In the lemma, let $f(a, b, c, d) = x^4 - (2a^2 + 2b^2 - a^2b^2)x^2 + (a^2 - b^2)^2$. At this time, we take the inscribed convex quadrilateral $ABCD$ of disk O of $(a, b, c, d) = (2(1 - v_0), 2(1 - 3v_0), 2(1 - 7v_0), 2(1 - 15v_0))$. As shown in Fig. 5.5, connect diagonal AC, let the perpendicular foot of B to AC be H_B and let the perpendicular foot of D to AC be H_D. With AB, BC, CD and DA as diameters, making disks O_{AB}, O_{BC}, O_{CD} and O_{DA}, triangles ABH_B, BCH_B, ADH_D and CDH_D are covered by disks O_{AB}, O_{BC}, O_{CD} and O_{DA}, respectively. Therefore, convex quadrilateral $ABCD$ is covered by these 4 disks. At this time, disk O must also be covered by disks O_{AB}, O_{BC}, O_{CD} and O_{DA}.

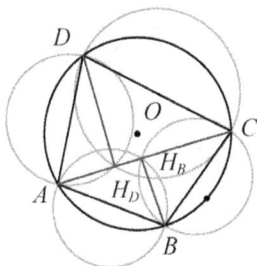

Fig. 5.5

It is known from the lemma that as long as it is proved that there exist $v_0 \in \left(\frac{1}{19}, \frac{1}{18}\right]$ and a permutation a, b, c, d of $2(1 - v_0), 2(1 - 3v_0),$ $2(1 - 7v_0), 2(1 - 15v_0)$ such that $f(a, b, c, d) > 0$.

In fact, let $v_0 = \frac{1}{19}, (a, b, c, d) = (2(1 - v_0), 2(1 - 3v_0), 2(1 - 7v_0),$ $2(1 - 15v_0))$, and then

$$(a, b, c, d) = \left(\frac{36}{19}, \frac{32}{19}, \frac{24}{19}, \frac{8}{19}\right) = \frac{4}{19}(9, 8, 6, 2),$$

$$x^2 = \frac{12 \times (81 + 64) + 72 \times (36 + 4)}{12 + 72} = 55 \times \left(\frac{4}{19}\right)^2,$$

so $f(a, b, c, d) = \left(\frac{4}{19}\right)^4 [55^2 - 55 \times (2 \times 81 + 2 \times 64 - 81 \times 64) + 289] > 0$. By the continuous sign-preserving property of $f(a, b, c, d)$, we know that there exists $v_0 \in \left(\frac{1}{19}, \frac{1}{18}\right]$ such that

$$f(2(1 - v_0), 2(1 - 3v_0), 2(1 - 7v_0), 2(1 - 15v_0)) > 0.$$

(2) We prove that $k = 18$ is the minimum value. That is, prove that when $v_0 \in \left(\frac{1}{18}, +\infty\right)$, John has no winning strategy. We use proof by contradiction.

If there exists $v_0 \in \left(\frac{1}{18}, +\infty\right]$, John still has a winning strategy. Then when $v_0 = \frac{1}{18}$, John has a winning strategy. Therefore, we can use at most 4 disks with radii $(1 - v_0), (1 - 3v_0), (1 - 7v_0), (1 - 15v_0)$, denoted as O_1, O_2, O_3, O_4 to cover disk O with radius 1.

We discuss the two cases as follows.

(A) As shown in Fig. 5.6, suppose the boundary of disk O with radius 1 intersects disks O_1, O_2, O_3, O_4 at arcs L_1, L_2, L_3, L_4, respectively, in clockwise order from the starting point. Then we must be able to find the inscribed convex quadrilateral $ABCD$ of disk O such that AB, BC, CD

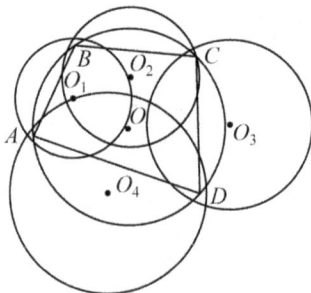

Fig. 5.6

and DA are contained within disks O_1, O_2, O_3, O_4, respectively. Therefore, the lengths of AB, BC, CD and DA must not exceed the diameters of
O_1, O_2, O_3 and O_4, respectively, that is, a certain permutation a, b, c, d of
$2(1 - v_0), 2(1 - 3v_0), 2(1 - 7v_0), 2(1 - 15v_0)$.

We know that the radius of disk O is an increasing function
of the side lengths AB, BC, CD and DA. By the lemma, there is
$f(AB, BC, CD, DA) \geqslant 0$. Therefore, it must hold for a permutation a, b, c, d
of $2(1 - v_0), 2(1 - 3v_0), 2(1 - 7v_0), 2(1 - 15v_0)$.

But in fact, when $v_0 = \frac{1}{18}$, fix $a = 2(1 - v_0)$, and no matter how b, c, d
are arranged, we find after calculation that $f(a, b, c, d) \geqslant 0$ does not hold.
Therefore, the assumption does not hold in this case.

(B) As shown in Fig. 5.7, there exists at least one disk $O_i, i \in \{1, 2, 3, 4\}$
such that this disk does not intersect disk O. Without loss of generality, we
can let it be O_4. We assume that the boundary of disk O intersects disk
O_1 at arc L_1. Since the radius of O_1 is smaller than that of O, there must
be two antipodal points E and F of O such that they are not covered by
disk O_1, and each of other disks O_2, O_3 can cover at most one of E, F (it
has been assumed that E and F are not within O_4). Therefore, O_2 and
O_3 must intersect disk O, which is set to arcs L_2, L_3. Let us assume that
L_1, L_2, L_3 be arranged in clockwise order from the starting point.

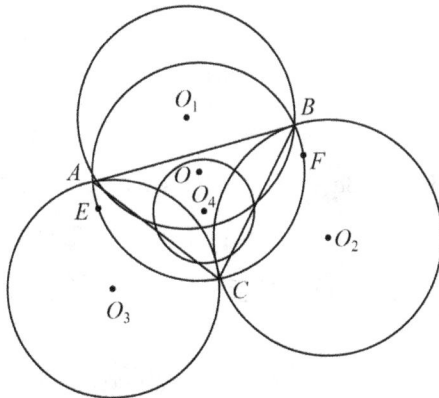

Fig. 5.7

Therefore, we must be able to find three clockwise points A, B, C on the
boundary of disk O such that AB, BC and CA are within disks O_1, O_2 and
O_3, respectively. Therefore, the lengths of AB, BC, CA must not exceed

the diameters of O_1, O_2, O_3, respectively. In addition, let $d = 0$, that is, D coincides with C (degenerating to a convex quadrilateral).

We know that the radius of disk O is an increasing function of the side lengths AB, BC, CD, DA. By the lemma, there is $f(AB, BC, CD, DA) \geqslant 0$. Therefore, it must hold for a permutation a, b, c, d of $2(1 - v_0), 2(1 - 3v_0), 2(1 - 7v_0), 2(1 - 15v_0)$.

But in fact, when $v_0 = \frac{1}{18}$, fix $a = 2(1 - v_0)$, and no matter how b, c, d are arranged, we find after calculation that $f(a, b, c, d) \geqslant 0$ does not hold. Therefore, the assumption does not hold in this case.

Therefore, the minimum value of k is 18.

11th Grade
First Day
(8:00–12:00; 2nd August 2022)

1 Let the real coefficient equation $x^3 + ax^2 + bx + c = 0$ in x have three positive real roots, and the sum of the three roots be not greater than 1.

Prove that $a^3(1 + a + b) - 9c(3 + 3a + a^2) \leqslant 0$.

(Contributed by Dong Qiuxian)

Solution 1 Assume that the three positive roots of the equation are x_1, x_2, x_3. Thus,

$$x_1 + x_2 + x_3 = -a, x_1 x_2 + x_2 x_3 + x_3 x_1 = b, x_1 x_2 x_3 = -c.$$

And

$$a^3(1 + a + b) - 9c(3 + 3a + a^2) = a^3(1 + a) + a^3 b - 27c(1 + a) - 9ca^2$$

$$= (1 + a)(a^3 - 27c) + a^2(ab - 9c).$$

Since $(x_1 + x_2 + x_3)^3 \geqslant (3\sqrt[3]{x_1 x_2 x_3})^3 = 27 x_1 x_2 x_3$, $-a^3 \geqslant -27c$, that is, $a^3 - 27c \leqslant 0$. And since

$$(x_1 + x_2 + x_3)(x_1 x_2 + x_2 x_3 + x_3 x_1) \geqslant 3\sqrt[3]{x_1 x_2 x_3} \cdot 3\sqrt[3]{x_1 x_2 \cdot x_2 x_3 \cdot x_3 x_1}$$

$$= 9 x_1 x_2 x_3,$$

$(-a)b \geqslant -9c$, namely, $ab - 9c \leqslant 0$. By the given $0 < -a \leqslant 1$, we have

$$(1 + a)(a^3 - 27c) + a^2(ab - 9c) \leqslant 0,$$

i.e., $a^3(1 + a + b) - 9c(3 + 3a + a^2) \leqslant 0$.

Solution 2 Assume that the three positive roots of the equation are x_1, x_2, x_3.

Given that at least one root of the equation is no greater than $\frac{1}{2}$, we might as well assume $x_3 \leqslant \frac{1}{2}$.

Since $x_1 x_2 \leqslant \frac{(x_1+x_2)^2}{4}$, we have $\frac{x_1 x_2}{1-x_1-x_2} \leqslant \frac{(x_1+x_2)^2}{4-4(x_1+x_2)}$, and then $\frac{x_1 x_2}{(1-x_1)(1-x_2)} \leqslant \frac{(x_1+x_2)^2}{[2-(x_1+x_2)]^2}$.

According to the KyFan inequality,

$$\frac{x_1 x_2 x_3}{(1-x_1)(1-x_2)(1-x_3)} \leqslant \frac{\left(\frac{x_1+x_2}{2}\right)^2}{\left[1-\frac{(x_1+x_2)}{2}\right]^2} \cdot \frac{x_3}{1-x_3}$$

$$\leqslant \frac{(x_1+x_2+x_3)^3}{[3-(x_1+x_2+x_3)]^3},$$

and thus

$$\frac{x_1 x_2 x_3}{1-(x_1+x_2+x_3)+(x_1 x_2 + x_1 x_3 + x_2 x_3)}$$

$$\leqslant \frac{(x_1+x_2+x_3)^3}{27 - 27(x_1+x_2+x_3) + 9(x_1+x_2+x_3)^2}.$$

By substituting $x_1+x_2+x_3 = -a, x_1 x_2 + x_2 x_3 + x_3 x_1 = b, x_1 x_2 x_3 = -c$, from the given conditions we can get $c < 0, a < 0$, and then $\frac{c}{1+a+b} \geqslant \frac{a^3}{9(3+3a+a^2)}$, which implies

$$a^3(1+a+b) - 9c(3+3a+a^2) \leqslant 0.$$

2 This question is the same as the question 2 for Grade 10.

3 There are n people standing in a queue, counting $1, 2, \ldots, n$ from left to right. Those who count odd numbers quit the queue. The remaining people count again from right to left, and those who count odd numbers quit the queue. The remaining people then count from left to right again. This process repeats until only one person remains in the queue. Let $f(n)$ represent the number of the last remaining person when he counted for the first time. Find the expression for $f(n)$ and determine the value of $f(2022)$.

(Contributed by Tao Pingsheng)

Solution From the problem statement, we know $f(1) = 1, f(2) = 2, f(3) = 2, \ldots$.

It is obvious that for any positive integer m, there is

$$f(2m+1) = f(2m). \qquad ①$$

Therefore, we only need to discuss the case where n is an even number.

When $n = 2m$, the beginning number of the last remaining person is $f(2m)$, and after the second count, this person should be at the $f(m)$th position from the right because there are only m people in total in the second count.

Analyze the relationship between two consecutive counts.

Let the kth person in the second count be at position x_k in the first count, and then
$$x_1 = 2m, x_2 = 2m - 2, x_3 = 2m - 4, \ldots, x_k = 2m - 2(k-1), \ldots,$$
$x_m = 2m - 2(m-1) = 2$. Therefore, $x_{f(m)} = 2m - 2(f(m) - 1)$, namely,

$$f(2m) = 2m + 2 - 2f(m). \qquad ②$$

From ②, we get

$$f(4m) = 4m + 2 - 2f(2m) = 4f(m) - 2. \qquad ③$$

And,

$$f(4m+2) = 2(2m+1) + 2 - 2f(2m+1)$$
$$= 4m + 4 - 2f(2m) = 4f(m). \qquad ④$$

Utilizing binary representation, the expression for $f(n)$ can be established. According to the above formulas ① and ②, we can combine them and get

$$f(n) = 2\left\lfloor\frac{n}{2}\right\rfloor + 2 - 2f\left(\left\lfloor\frac{n}{2}\right\rfloor\right).$$

Suppose the binary expression of the number n is $n = (\overline{a_k a_{k-1} \cdots a_2 a_1})$, $a_k = 1, a_i \in \{0,1\}, i < k$, and then

$$f(n) = 2(\overline{a_k a_{k-1} \cdots a_3 a_2}) + 2 - 2f(\overline{a_k a_{k-1} \cdots a_3 a_2})$$
$$= 2(\overline{a_k a_{k-1} \cdots a_3 a_2}) + 2 - 2\left[2(\overline{a_k a_{k-1} \cdots a_3}) + 2 - 2f(\overline{a_k a_{k-1} \cdots a_3})\right]$$
$$= \overline{a_k a_{k-1} \cdots a_2 0} + 2 - \overline{a_k a_{k-1} \cdots a_3 00} - 4 + 4f(\overline{a_k a_{k-1} \cdots a_3})$$
$$= 2(a_2 - 1) + 4f(\overline{a_k a_{k-1} \cdots a_3}),$$

namely, $f(\overline{a_k a_{k-1} \cdots a_2 a_1}) = 2(a_2 - 1) + 4f(\overline{a_k a_{k-1} \cdots a_3})$.

Similarly, there are

$$f(\overline{a_k a_{k-1} \cdots a_3}) = 2(a_4 - 1) + 4f(\overline{a_k a_{k-1} \cdots a_5}),$$

$$f(\overline{a_k a_{k-1} \cdots a_5}) = 2(a_6 - 1) + 4f(\overline{a_k a_{k-1} \cdots a_7}),$$

etc.

Therefore, when $n = (\overline{a_{2m+1} a_{2m} \cdots a_2 a_1}), a_{2m+1} = 1, a_i \in \{0, 1\}, i < 2m+1$, that is, when the binary representation of n contains an odd number of digits, we have

$$f(n) = 2(a_2 - 1) + 2^3(a_4 - 1) + \cdots + 2^{2m-1}(a_{2m} - 1) + 2^{2m} f(a_{2m+1})$$

$$= 2(a_2 - 1) + 2^3(a_4 - 1) + \cdots + 2^{2m-1}(a_{2m} - 1) + 2^{2m}.$$

When $n = (\overline{a_{2m+2} a_{2m+1} \cdots a_2 a_1}), a_{2m+2} = 1, a_i \in \{0, 1\}, i < 2m + 2$, that is, when the binary representation of n contains an even number of digits, we have

$$f(n) = 2(a_2 - 1) + 2^3(a_4 - 1) + \cdots + 2^{2m-1}(a_{2m} - 1)$$

$$+ 2^{2m} f(\overline{a_{2m+2} a_{2m+1}})$$

$$= 2(a_2 - 1) + 2^3(a_4 - 1) + \cdots + 2^{2m-1}(a_{2m} - 1) + 2^{2m+1}.$$

In particular, by $2022 = (11111100110) = 2^{10} + 2^9 + 2^8 + 2^7 + 2^6 + 2^5 + 2^2 + 2$, it follows that

$$f(2022) = 2(1 - 1) + 2^3(0 - 1) + 2^5(1 - 1) + 2^7(1 - 1) + 2^9(1 - 1) + 2^{10}$$

$$= 1024 - 8 = 1016.$$

4 If x_i is a positive integer greater than 1, denote $f(x_i)$ as the largest prime factor of x_i. Let $x_{i+1} = x_i - f(x_i)$, where i is a natural number.

(1) Prove that for any positive integer greater than 1, there exists a natural number k such that $x_{k(x_0)+1} = 0$;

(2) let $V(x_0)$ be the number of distinct values in $f(x_0), f(x_1), \ldots, f(x_{k(x_0)})$. Determine the maximum number in $V(2), V(3), \ldots, V(2022)$ and give your reasons.

(Contributed by Luo Ye)

Solution Let $p_i = f(x_i)$ be a prime number.

(1) By definition, for any i, if $x_i \in \mathbf{N}_+, x_i > 1$, it is easy to see that $x_i > x_{i+1} \geqslant 0$ is a descending integer sequence and $p_i \mid x_{i+1}$. Therefore, for any initial $x_0 \in \mathbf{N}_+, x_0 > 1$, this descent can only be performed a finite number of times. Therefore, there must exist a natural number k such that

$x_k \geqslant 2, x_{k+1} < 2$. But since $x_{k+1} \geqslant 0$ and $p_k \mid x_{k+1}, x_{k+1}$ cannot be 1, there must be $x_{k+1} = 0$.

(2) It is easy to see that p_i is an increasing sequence, while x_i is a strictly decreasing sequence, $i = 0, 1, \ldots, k$. Let $i_1, \ldots, i_s$ be the number in $1, \ldots, k$ such that $p_{i_l} > p_{i_l - 1}, l = 1, 2, \ldots, s$. Thus, $V(x) = s + 1$.

Let $q_0 = p_0, q_1 = p_{i_l}, l = 1, 2, \ldots, s$. But $q_0, \ldots, q_s$ are all different prime numbers in $f(x_0), \ldots, f(x_k)$, and $q_s > q_{s-1} > \cdots > q_0$. It is easy to find $q_l q_{l-1} \mid x_{i_l}, l = 1, 2, \ldots, s$. Let $k_l = \frac{x_{i_l}}{q_l q_{l-1}}$. It can be seen that $k_s < k_{s-1} < \cdots < k_1$ is a strictly decreasing sequence of positive integers. First we prove $s \leqslant 4$. Now inversely assume $s \geqslant 5$, namely, $V(x_0) \geqslant 6$.

So there is $q_5 \geqslant 13, q_4 \geqslant 14$, and then $x_0 > 13 \times 11 = 143$. If $q_0 = 2$, then x_0 is a power of 2. Since $x_0 \leqslant 2022, x_0 \in \{256, 512, 1024\}$. But it can be verified that $V(256) = 2, V(512) = 3, V(1024) = 2$, contradicting $V(x_0) = 6$. Similarly, $q_0 \neq 3$ can be verified. Therefore, $q_0 \geqslant 5$, and thus $q_5 \geqslant 19, q_4 \geqslant 17, q_3 \geqslant 13, q_2 \geqslant 11, q_1 \geqslant 7$. Thus, we have $x_0 > 17 \times 19 = 323$, so $k_5 \leqslant 6, k_4 \leqslant 9$.

It can be confirmed that $q_{s-1} < 43$, otherwise there is $x_0 \geqslant V(x_0) - 1 + x_{i_s} \geqslant 5 + q_{s-1} q_s \geqslant 5 + 43 \times 47 = 2026$, a contradiction. Therefore, $q_{s-1} \leqslant 41$, i.e., we have $q_4 \leqslant 41$.

By definition, we know that $x_{i_l}, x_{i_l+1}, \ldots, x_{i_{l+1}}$ is an arithmetic sequence and covers all multiples of q_l from $q_l(q_{l+1}k_{l+1})$ to $q_l(q_{l-1}k_l)$. Thus, we have the following properties:

Property (i): $q_{l+1}k_{l+1} < q_{l-1}k_l$ and the maximum prime factor of every number in $q_{l+1}k_{l+1} + 1, \ldots, q_{l-1}k_l$ is not greater than q_l.

Property (ii):   There is at least one prime number for every 11 consecutive numbers between 1 and 118.

Property (iii):   If $q_{l-1} \geqslant 11$ and $q_{l-1}k_l \leqslant 118$, then $q_{l-1}k_l$ is exactly the first number greater than or equal to $q_{l+1}k_{l+1} + 1$ and divisible by q_{l-1}. Specifically, define $I_{l+1} = \{q_{l+1}k_{l+1} + 1, q_{l+1}k_{l+1} + 2, \ldots, q_{l-1}k_l\}, l = 1, 2, \ldots, s - 1$.

Then by properties (i) and (ii), we know that none of the numbers in I_{l+1} can be prime numbers, nor can they have prime factors larger than q_l.

In the following we prove $q_2 = 11$.

If $q_2 \geqslant 13$, then $q_3 \geqslant 17, q_4 \geqslant 19$. We have $k_4 q_4 < \frac{x_0}{q_3}$, so $k_4 q_4 \leqslant 118$. But q_4 can only take one of the values in $19, 23, 29, 31, 37, 41$. Next, we discuss the different cases.

(a) $q_4 = 19$, then $2 \leqslant k_4 \leqslant 6, q_2 = 13, q_3 = 17$. If $k_4 = 2$, then $k_4 q_4 + 1 = 39 \in I_4$ is exactly the multiple of 13. By property (iii), there is $k_3 = 3$.

Thus, $x_{l_3} = 3 \times 13 \times 17$. However, $k_3 q_3 + 1 = 52$ cannot be divisible by q_1, while $k_3 q_3 + 2 = 53 \in I_3$ is a prime number, contradicting property (iii) (hereinafter referred to as a contradiction)! If $k_4 = 3$, then $k_4 q_4 + 1 = 58 \in I_4$ has a prime factor $29 > q_3$, a contradiction. If $k_4 = 4$, then $k_4 q_4 + 1 = 77$, and $k_4 q_4 + 2 = 78$ is a multiple of 13. Thus, by property (iii) we have $k_3 = 6$. But $k_3 q_3 + 1 = 6 \times 17 + 1 = 103 \in I_3$ is a prime number, a contradiction. If $k_4 = 5$, then $k_4 q_4 + 2 = 97 \in I_4$ is a prime number, a contradiction. If $k_4 = 6$, then $k_4 q_4 + 1 = 115 \in I_4$ is a multiple of 23, a contradiction.

(b) $q_4 = 23$, then $2 \leqslant k_4 \leqslant 5$. If $k_4 = 2$, then $k_4 q_4 + 1 = 47 \in I_4$, a contradiction. If $k_4 = 3$, since $k_4 q_4 + 1 = 70$ is not a multiple of $q_2 \geqslant 13$, then $k_4 q_4 + 2 = 71 \in I_4$, a contradiction. If $k_4 = 4$, then $k_4 q_4 + 1 = 93$ is a multiple of 31, a contradiction. If $k_4 = 5$, then $k_4 q_4 + 1 = 116$ is a multiple of 29, a contradiction.

(c) $q_4 = 29$, then $2 \leqslant k_4 \leqslant 4$. If $k_4 = 2$, then $k_4 q_4 + 1 = 59$ is a prime number, a contradiction. If $k_4 = 3$, then $k_4 q_4 + 1 = 88$ has no prime factor less than 13, and hence $k_4 q_4 + 2 = 89$ is a prime number, a contradiction. If $k_4 = 4$, then $k_4 q_4 + 1 = 117 = 9 \times 13$; while 118 is a multiple of 59, so $q_2 = 13, k_3 = 9$; and since $q_3 \leqslant \frac{2022}{29 k_4} \leqslant 17.5$, $q_3 = 17$. Since $k_3 q_3 + 1 = 154$ is a multiple of 11 and $k_3 q_3 + 2 = 155$ can be divisible by 31, there must be $q_1 = 11, k_2 = 14$; however, $k_2 q_2 + 1 = 183$ is a multiple of 61, a contradiction.

(d) $q_4 = 31$, then $2 \leqslant k_4 \leqslant 3$. If $k_4 = 2$, then $k_4 q_4 + 1 = 63$ cannot be a multiple of $q_2 \geqslant 13$. Similarly, 64 and 66 are also not . If $q_2 > 13$, then 65 is not, and then $67 \in I_4$, a contradiction. Therefore, $q_2 = 13, k_3 = 5$. At this time the possible values of q_3 are $17, 19, 23, 29$. But $17 \times 5 + 1 = 86$ is a multiple of 43, $19 \times 5 + 1 = 96$ is not a multiple of q_1, $19 \times 5 + 2 = 97$ is a prime number, $23 \times 5 + 1 = 116$ is a multiple of 29, and $29 \times 5 + 1 = 146$ is a multiple of 73, all of which are not possible. If $k_4 = 3$, then $k_4 q_4 + 1 = 94$ is a multiple of 47, a contradiction.

(e) $q_4 = 37$, then $2 \leqslant k_4 \leqslant 3$. If $k_4 = 2$, then $k_4 q_4 + 1 = 75$. Since 75 and 77 have no prime factors less than 13 and 79 is a prime number, (1) $q_2 = 19, k_3 = 4$, or (2) $q_2 = 13, k_3 = 6$ and $q_3 \geqslant 19$.

    In Case 1, $q_3 < \frac{2022}{q_4 k_4} < 29$, so $q_3 = 23$. But $k_3 q_3 + 1 = 113 \in I_3$, a contradiction.

    In Case 2, q_3 is also less than 29, so $q_3 = 19$ or $q_3 = 23$. If $q_3 = 19$, then $k_3 q_3 + 1 = 115$ is a multiple of 23, a contradiction. If $q_3 = 23$, then $k_3 q_3 + 1 = 139$ is a prime number, a contradiction.

    If $k_4 = 3$, then $k_4 q_4 + 1 = 112$ is not a multiple of $q_3 \geqslant 13$, while $k_4 q_4 + 2 = 113$ is a prime number, a contradiction.

(f) $q_4 = 41$, then $k_4 = 2$. Then $k_4 q_4 + 1 = 83$ is a prime number, a contradiction.

Thus, q_2 must be 11, and then $q_0 = 5, q_1 = 7$. Therefore, $4 \leqslant k_2 \leqslant \frac{2022}{q_1 q_2} < 27$. By property (iii), we can first exclude those values of k_2 such that $11 k_2 + 1$ has a prime number not less than 13. So the possible values of k_2 are $4, 9, 13, 19, 22$. But when $k_2 = 13$ or $k_2 = 22$, although $k_2 q_2 + 1$ does not have a prime factor greater than or equal to 13, it is not a multiple of $q_0 = 5$, and $k_2 q_2 + 2$ has a prime factor not less than 13. Thus, only the cases of $k_2 = 4, 9, 19$ need to be considered. Considering $V(x_0) \geqslant 6$, we have $V(x_0) \geqslant 4$. When $k_2 = 4, x_{i_2} = 4 \times 7 \times 11 = 308, V(x_{i_2}) = 3 < 4$ (*Note*: there must be $V(x_0) = 5$ at this time). When $k_2 = 9, x_{i_2} = 9 \times 7 \times 11, V(x_{i_2}) = 2$. When $k_2 = 19, x_{i_2} = 19 \times 7 \times 11$. But at this time, $f(x_{i_2}) = 19 = q_2$, while $q_2 = 11$, a contradiction.

In conclusion, $\max_{2 \leqslant x_0 \leqslant 2022} V(x_0) = 5$. And this number can be obtained exactly when $x_0 = 320$.

11th Grade
Second Day
(8:00–12:00; 3rd August 2022)

⑤ Let a, b, c, d be non-negative integers, and $S = a + b + c + d$.

(1) If $a^2 + b^2 - cd^2 = 2022$, find the minimum value of S;
(2) If $a^2 - b^2 + cd^2 = 2022$, find the minimum value of S.

(Contributed by He Yijie and Zhang Hongshen)

Solution (1) First of all, c and d are both non-zero (otherwise there will be $a^2 + b^2 = 2022$, and then $a^2 + b^2 \equiv 0 \pmod 3$, so a, b will be multiples of 3, leading to $9 \mid a^2 + b^2$, which contradicts $a^2 + b^2 = 2022$).

Since $(a + b)^2 \geqslant a^2 + b^2 \geqslant 2022$, $a + b \geqslant 45$.

If $a + b = 45$, it can only be $\{a, b\} = \{0, 45\}$ (otherwise $a^2 + b^2 \leqslant 1^2 + 44^2 < 2022$, a contradiction). At this time, $cd^2 = 0^2 + 45^2 - 2022 = 3$, which gives $c = 3, d = 1$, and correspondingly there is $S = 49$.

If $a + b = 46$, a and b are of the same parity, and thus cd^2 is an even number. Therefore, c and d contain even numbers, and $c + d$ is at least 3, giving $S \geqslant 46 + 3 = 49$ (when $\{a, b\} = \{1, 45\}, c = 1, d = 2$, the equality can be reached).

If $a + b \geqslant 47$, then $S \geqslant 47 + 1 + 1 \geqslant 49$.

In summary, the minimum value of S is 49.

(2) First of all, there is $0^2 - 1^2 + 7 \times 17^2 = 2022$, so $(a, b, c, d) = (0, 1, 7, 17)$ satisfies the condition, giving $S = 25$.

In the following we assume $S \leqslant 24$.

If $c + d = 24$, then $a = b = 0$. At this time, $cd^2 = 2022$. Note that 2022 has no square factors, and thus $d = 1, c = 2022$, which contradicts $c + d = 24$.

If $c + d \leqslant 23$, by the AM-GM inequality, we have

$$cd^2 = 4c \cdot \frac{d}{2} \cdot \frac{d}{2} \leqslant 4 \cdot \left[\frac{1}{3} \left(c + \frac{d}{2} + \frac{d}{2} \right) \right]^3 = \frac{4}{27}(c+d)^3. \qquad ①$$

Therefore,

$$cd^2 \leqslant \frac{4}{27} \times 23^3 < \frac{7}{2} \times 23^2 < 1900.$$

At this point, $a^2 = 2022 + b^2 - cd^2 > 2022 - 1900 = 122$, so $a > 11$.

Thus, $c + d = S - a - b \leqslant 24 - a < 13$. Then by using ①, we get

$$cd^2 \leqslant \frac{4}{27} \times 13^3 < 2 \times 13^2 < 400.$$

Hence, $a^2 > 2022 - 400 > 1600$, giving $a > 40 > S$, a contradiction.

In summary, the minimum value of S is 25.

Remark In estimating the lower bound of S in problem 2, the inequality ① plays an important role. In fact, if cd^2 is larger, it is known from ① that $c + d$ is also larger, leading to a larger S; if cd^2 is smaller, then it is known from the condition that $a = \sqrt{2022 + b^2 - cd^2}$ is larger, also leading to a larger S.

There are many specific methods of estimating the lower bound and presenting the answer, for example, one of which is also shown in the following.

Assume $S \leqslant 24$, and then $a \leqslant 24$, so $cd^2 = 2022 + b^2 - a^2 \geqslant 2022 - 24^2 = 1446$. By the arithmetic mean-geometric mean inequality, we have

$c + d = c + \frac{d}{2} + \frac{d}{2} \geqslant 3\sqrt[3]{\frac{cd^2}{4}} > 21$ (an equivalent form of ①),

so $c + d \geqslant 22$. Therefore,

$$S^3 \geqslant 3(a + b)^2 \cdot (c + d) + 3(a + b) \cdot (c + d)^2 + (c + d)^3$$

$$\geqslant 3(a + b)^2 \cdot 22 + 3(a + b) \cdot 22^2 + \frac{27}{4}cd^2$$

$$\geq \frac{27}{4}(a^2 - b^2) + 3(a+b) \cdot 22^2 + \frac{27}{4}cd^2$$

$$= \frac{27}{4} \cdot 2022 + 3(a+b) \cdot 22^2.$$

If $a + b \geq 1$, then $S^3 \geq \frac{27}{4} \cdot 2022 + 3 \cdot 22^2 > 24^3$, contradicting $S \leq 24$.

Thus, $a + b = 0$, and then $cd^2 = 2022$. Note that 2022 has no square factors, and thus $d = 1, c = 2022$. At this time, $S > 24$, a contradiction.

When identifying whether there is substantial progress in the lower bound estimation of S in question 2, 'the occurrence of inequality ① or its equivalent form' may serve as a major observation point.

⑥   As shown in Fig. 5.8, point O is the circumcenter of acute-angled $\triangle ABC$. $\odot P$ passes through points A, O, and $OP \parallel BC$. Points D and A are on opposite sides of side BC, satisfying $\angle ABD = \angle ACD = \angle BAC$. $\odot Q$ is the circle with AD as its diameter, and $\odot R$ is the circumcircle of $\triangle BCD$. Prove that $\odot P$, $\odot Q$ and $\odot R$ intersect at the same point.

(Contributed by Lin Tianqi)

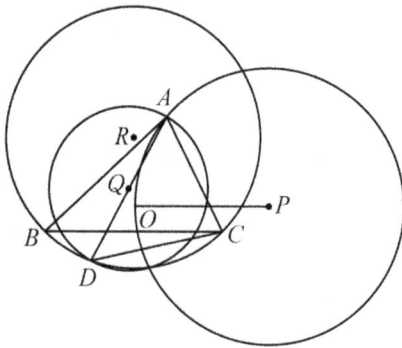

Fig. 5.8

Solution   As shown in Fig. 5.9, let the extensions of AC and BD intersect at point U, the extensions of AB and CD intersect at point V. Suppose $\odot R$ and the circumcircle of $\triangle UDV$ intersect at points D and M. In the following we will prove that M is the common point of $\odot P$, $\odot Q$ and $\odot R$.

Taking the symmetry point A' of A about OP, then A' is an intersection of $\odot Q$ and $\odot P$, and $AA' \perp BC$. Let N be the midpoint of BC.

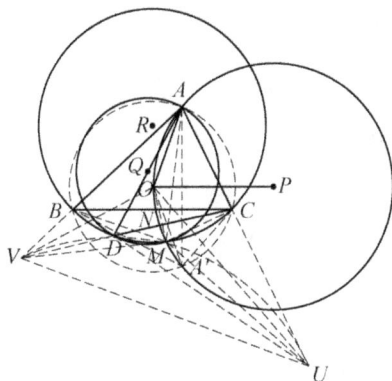

Fig. 5.9

Since $\angle ABD = \angle ACD = \angle BAC$, we have $AU = BU$ and $AV = CV$. And since O is the circumcenter of $\triangle ABC$, we get $OU \perp AB$ and $OV \perp AC$, and then O is the orthocenter of AUV. Therefore,

$$\angle OUV = \angle OAB = 90° - \angle ACB = 90° - \angle AA'B = \angle A'BC,$$

namely, $\angle OUV = \angle A'BC$. Similarly, we can get $\angle OVU = \angle A'CB$, and thus $\triangle OUV \backsim \triangle A'BC$.

By the inscribed angle theorem, we have

$$\angle MBC = \angle MDC = \angle MUV,$$

namely, $\angle MBC = \angle MUV$. Similarly, we can get $\angle MCB = \angle MVU$, and thus $\triangle MBC \backsim \triangle MUV$.

Combining $\triangle OUV \backsim \triangle A'BC$ and $\triangle MBC \backsim \triangle MUV$, we have

$$\angle A'MO = \angle(BC, UV) = \angle NOA,$$

namely, $\angle A'MO = \angle NOA$. And by $ON \parallel AA'$, it follows that

$$\angle A'MO + \angle OAA' = \angle NOA + \angle OAA' = 180°,$$

so points A, O, M, A' are concyclic, that is, M lies on $\odot P$.

And it is obvious that $\angle ABU = \angle ACV$, so points B, C, U, V are concyclic. By the properties of the Miquel point, we know that $AM \perp DM$, and thus M also lies on $\odot Q$.

In summary, $\odot P, \odot Q$ and $\odot R$ are concurrent at point M, and thus the conclusion is established.

7 Prove that for any real number $\lambda > 0$, there exist n positive integers $a_1, a_2, \ldots, a_n (n \geqslant 2)$, satisfying $a_1 < a_2 < \cdots < a_n < 2^n \lambda$, and for any $1, 2, \ldots, n$, there is

$$(a_1, a_k) + (a_2, a_k) + \cdots + (a_n, a_k) \equiv 0 (\mathrm{mod}\ a_k),$$

where (u, v) denotes the greatest common divisor of integers u and v.

(Contributed by He Yijie)

Solution Take a positive integer $r \geqslant 2$, satisfying $2^{2^r - r + 1} \lambda > 1$.

Consider the positive integers $n = b_1 b_2 \cdots b_{2^r} + 1$ and the n positive integers below

$$b_1, b_2, \ldots, b_{2^r}, b_{2^r + 1} = 2^r, b_{2^r + 2} = 2^{r+1}, \ldots, b_n = 2^{n - 2^r + r - 1} \qquad \text{①}$$

where $b_1 < b_2 < \cdots < b_{2^r}$ are all odd prime numbers.

For each $k(1 \leqslant k \leqslant 2^r)$, it is obvious that b_k are coprime with the remaining numbers in ①, and hence

$$(b_1, b_k) + (b_2, b_k) + \cdots + (b_n, b_k) = n - 1 + b_k$$

$$= b_1 b_2 \cdots b_{2^r} + b_k \equiv 0 (\mathrm{mod}\ b_k). \qquad \text{②}$$

For each $k(2^r + 1 \leqslant k \leqslant n)$, there is

$$(b_i, b_k) = \begin{cases} 1, & 1 \leqslant i \leqslant 2^r, \\ b_i, & 2^r + 1 \leqslant i \leqslant k - 1, \\ b_k, & k \leqslant i \leqslant n. \end{cases}$$

Therefore,

$$(b_1, b_k) + (b_2, b_k) + \cdots + (b_n, b_k) \equiv 2^r \cdot 1 + b_{2^r + 1} + b_{2^r + 2} + \cdots + b_{k-1}$$

$$= 2^r + 2^r + 2^{r+1} + \cdots + 2^{r + k - 2^r - 2} = 2^{r + k - 2^r - 1}$$

$$= b_k \equiv 0 (\mathrm{mod}\ b_k). \qquad \text{③}$$

It is obvious that $b_{2^r} > 2^r$. Thus,

$$n - 2^r + r - 1 = b_1 b_2 \cdots b_{2^r} - 2^r + r > 2 b_{2^r} - 2^r > b_{2^r},$$

and hence $b_n > 2^{b_2} > b_{2^r}$. Therefore, $b_n = \max\{b_1, b_2, \ldots, b_n\}$.

It is clear that $b_1, b_2, \ldots, b_n$ are distinct. Arrange $b_1, b_2, \cdots, b_n$ in ascending order as $a_1, a_2, \ldots, a_n$. Note that $2^{r^r - r + 1} \lambda > 1$, and there is

$$a_1 < a_2 < \cdots < a_n = b_n = 2^{n - 2^r + r - 1} < 2^n \lambda.$$

Combining ② and ③, it can be seen that n and $a_1, a_2, \ldots, a_n$ satisfy the conditions.

Remark The format and the proof idea of this problem were inspired by the 2018 Iran National Team Selection Test problem. The original problem is as follows:

Different positive integers $a_1, a_2, \ldots, a_n$ are said to be 'harmonic' if and only if

$$\sum_{i=1}^{n} a_i = \sum_{1 \leqslant i < j \leqslant n} (a_i, a_j).$$

Prove that there exist infinitely many positive integers n such that there are n positive integers that are harmonic.

One way to prove this problem is by taking $a_1, a_2, \ldots, a_n$ in the following form

$$a_1 = 2^0, a_2 = 2^1, \ldots, a_{n-2} = 2^{n-3}, a_{n-1} = 3, a_n = q \text{(where } (q, 6) = 1).$$

A simplification of

$$\sum_{i=1}^{n} a_i = \sum_{1 \leqslant i < j \leqslant n} (a_i, a_j)$$

shows that the stated equation holds when $n = q + 4$.

8 This question is the same as the question 8 for Grade 10.

6

China Mathematical Olympiad Hope Alliance Summer Camp

2022 (Shijiazhuang, Hebei)

The China Mathematical Olympiad "Hope Alliance" was established in 2001. It was proposed by Mr. Qiu Zonghu, the then-director of the Organizing Committee of the Chinese Mathematical Society. Its aim is to unite schools with outstanding performances in National Mathematics Competitions to promote further achievements and collaboration, specifically targeting the International Mathematical Olympiad (IMO). This is a communication platform jointly built to explore the rules of discovering and cultivating mathematical science talents. The "Hope Alliance" holds two major events each year. Principals and coaches' meetings are usually held in the second half of the year to summarize the league's operations in the previous year and the development direction and goals for the next few years; the summer camp that focuses on intensive training, lectures by distinguished teachers, problem-solving sessions, and exchange activities among member school students. The camp is designed to enhance the communication and interaction between students from different schools, improve mathematical understanding, and raise the overall teaching and student performance levels. From 2002, when Hope Alliance member schools won the first IMO gold medal, to 2022, eight schools with 14 IMO national team members have won 12 gold and two silver medals. This year, a student from another school successfully entered the IMO National Team. Currently, the Alliance has 16 member schools: Attached Middle School to Jilin University, Liaoning

Province Shiyan High School, Lanzhou No.1 Senior High School, Beijing National Day School, Shijiazhuang No. 2 High School in Hebei Province, Xi'an Gaoxin No. 1 High School, Affiliated Middle School of Henan Normal University, Jiangsu Xishan Senior High School, Zhenhai High School in Zhejiang Province, Nanchang No. 2 Middle School in Jiangxi Province, Yali Middle School in Hunan Province, Xiamen Shuangshi Middle School of Fujian, Chengdu Jiaxiang Foreign Languages School, Chongqing Bashu Secondary School, Shandong Experimental High School, Suzhou High School of Jiangsu Province. In 2022, Tianjin Nankai High School, Guangdong Sun Yat-sen Memorial Secondary School, and Tongliao No. 5 Senior High School in Inner Mongolia participated in the program as observing schools.

The 2022 "Hope Alliance" Summer Camp, hosted by Shijiazhuang No. 2 High School in Hebei Province, had three tests in total. The first and third tests included the first and second rounds of the China Mathematical Competition, and the second test was slightly higher than the second round of the China Mathematical Competition. The following are the questions and answers from the second test of this year's "Hope Alliance" summer camp.

8:30–11:20; 28th July 2022

Find all integers n such that for any three distinct positive real numbers a, b, c, there is

$$\frac{a^{n+1}}{|b-c|^n} + \frac{b^{n+1}}{|c-a|^n} + \frac{c^{n+1}}{|a-b|^n} \geqslant a+b+c.$$

Solution n are natural numbers.

When $n = 0$, it is obvious.

When $n \geqslant 1$, without loss of generality, let $c = \min\{a, b, c\}$. By the AM-GM inequality, we have

$$\frac{a^{1+n}}{|b-c|^n} + n|b-c| \geqslant (1+n)a,$$

$$\frac{b^{1+n}}{|c-a|^n} + n|c-a| \geqslant (1+n)b,$$

$$\frac{a^{1+n}}{|b-c|^n} + \frac{b^{1+n}}{|c-a|^n} + \frac{c^{1+n}}{|a-b|^n}$$

$$\geqslant (1+n)a + (1+n)b - n|b-c| - n|c-a| = a+b+2nc \geqslant a+b+c.$$

When $n < 0$, take $a = 1 + \frac{1}{m}, b = 1, c = 1 - \frac{1}{m}$ and let $m \to +\infty$, and then $0 > 3$, a contradiction.

2 Given that sequence $\{F_n\}$ satisfies $F_0 = 0, F_1 = 1, F_{n+2} = F_{+1} + F_n (n \in \mathbf{N})$. For an integer m greater than 2, denote R_m as the remainder of $\prod_{k=1}^{F_m-1} k^k$ divided by F_m. Prove that R_m is in sequence $\{F_n\}$.

Solution From $m > 2$, we know that $F_m \geqslant 2$. When F_m is a composite number, $F_m \mid \prod_{k=1}^{F_m-1} k^k$. So $R_m = 0 = F_0$, and the conclusion holds.

When F_m is a prime number, if $F_m = 2$, then $R_m = 1 = F_1$, and the conclusion holds.

When F_m is an odd prime p, then

$$\prod_{k=1}^{p-1} k^k = \prod_{k=1}^{\frac{p-1}{2}} k^k (p-k)^{p-k} \equiv \prod_{k=1}^{\frac{p-1}{2}} k^p (-1)^{p-k} \equiv \prod_{k=1}^{\frac{p-1}{2}} k(-1)^{k+1}$$

$$= \left(\frac{p-1}{2}\right)!(-1)^{\frac{(p-1)(p+5)}{8}} \pmod{p}.$$

By $F_0 = 0, F_1 = 1, F_{n+2} = F_{n+1} + F_n (n \in \mathbf{N})$, it follows that

$$F_n = \frac{\sqrt{5}}{5}\left[\left(\frac{1+\sqrt{5}}{2}\right)^n - \left(\frac{1-\sqrt{5}}{2}\right)^n\right].$$

And there is

$$F_{m+1} \cdot F_{m-1} - F_m^2 = \frac{1}{5}\left[\left(\frac{1+\sqrt{5}}{2}\right)^{m+1} - \left(\frac{1-\sqrt{5}}{2}\right)^{m+1}\right]$$

$$\times \left[\left(\frac{1+\sqrt{5}}{2}\right)^{m-1} - \left(\frac{1-\sqrt{5}}{2}\right)^{m-1}\right]$$

$$- \frac{1}{5}\left[\left(\frac{1+\sqrt{5}}{2}\right)^m - \left(\frac{1-\sqrt{5}}{2}\right)^m\right]^2$$

$$= (-1)^m.$$

Thus, for $2 \nmid m$, $F_{m-1}^2 \equiv -1 \pmod{F_m}$, so $F_{m-1}^{p-1} \equiv (-1)^{\frac{p-1}{2}} \pmod{p}$. However, by Fermat's little theorem, $F_{m-1}^{p-1} \equiv 1 \pmod 4$, hence, $p \equiv 1 \pmod 4$.

By the general term, we get $F_m \mid F_{mn} (m, n \in \mathbf{Z}_+)$. So by $F_m = p$, we know that m is a prime number or $m \leqslant 4$.

(1) If m is a prime number, then $2 \nmid m$, $F_m = p \equiv 1 \pmod 4$.

And $R_m \equiv \left(\frac{p-1}{2}\right)!(-1)^{\frac{(p-1)(p+5)}{8}} \pmod p$, so by Wilson's theorem $(p-1)! \equiv -1 \pmod p$, there is

$$R_m^2 \equiv \left[\left(\frac{p-1}{2}\right)!\right]^2 \equiv (-1)^{\frac{p-1}{2}} \cdot (p-1)! \equiv (-1)^{\frac{p+1}{2}} \equiv -1 \pmod p.$$

Hence, $R_m \equiv F_{m-1}$ or $R_m \equiv -F_{m-1} \pmod{F_m}$. Therefore, $R_m \equiv F_{m-1}$ or $R_m \equiv F_{m-2}$.

(2) If $m \leqslant 4$, then the solutions of $F_m = p$ are only $F_4 = 3$, $\prod_{k=1}^{F_m-1} k^k = 4$, $R_4 = 1 = F_1$, and the conclusion holds.

The proof is completed.

Remark Some results are not obvious. To prove $F_m \mid F_{mn}(m, n \in \mathbf{Z}_+)$, we can first prove that $F_n = \left(\frac{1+\sqrt 5}{2}\right)^n - \left(\frac{1-\sqrt 5}{2}\right)^n$ is the general term of sequence $F_n = \left(\frac{1+\sqrt 5}{2}\right)^n - \left(\frac{1-\sqrt 5}{2}\right)^n$.

8 As shown in Fig. 6.1, in quadrilateral $ABCD$ inscribed in a circle, $BD > AC$. Lines BC and AD intersect at E, and lines AB and CD intersect at F. Draw a line through E parallel to line CD, intersecting line AB at K. U and V are the circumcenters of $\triangle FAC$ and $\triangle FBD$, respectively.

Prove that $\angle UKV = \angle BCD - \angle ABC$.

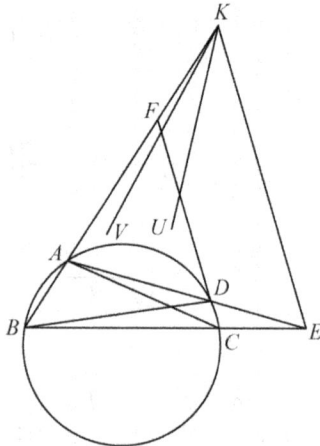

Fig. 6.1

Solution 1   As shown in Fig. 6.2, draw a line through E parallel to AB that intersects line CD at T.

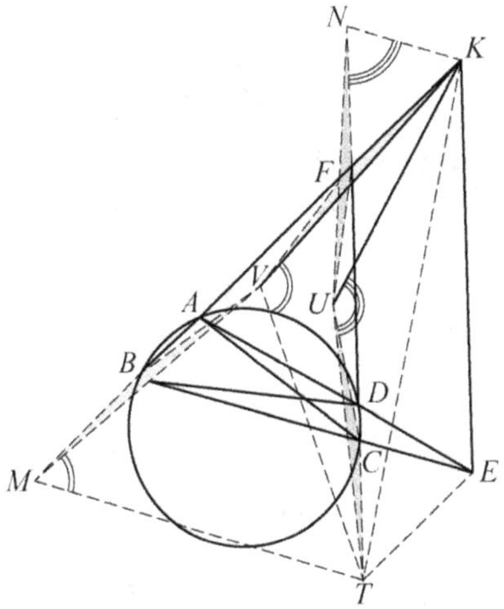

Fig. 6.2

We have

$$\frac{KB \cdot KF}{EB \cdot EC} = \frac{FB^2}{BC^2} = \frac{FD^2}{DA^2} = \frac{TD \cdot TF}{ED \cdot EA}.$$

Therefore, $KB \cdot KF = TD \cdot TF$.

Thus, K and T have equal powers to the circumcircle of $\triangle BDF$, so $VK = VT$. Similarly, $UK = UT$.

By the fact that $KFTE$ is a parallelogram, draw the symmetry point M of K about the midpoint of BF, and the symmetry point N of T about the midpoint of CF.

By symmetry, we know that $VM = VK, BM = KF = TE$. Therefore, quadrilateral $BMTE$ is a parallelogram, and V is the circumcenter of $\triangle KMT$. Therefore, $MT \parallel BE$, and then $\angle KMT = \angle ABC$. Hence, $\angle VKT = 90° - \angle KMT = 90° - \angle ABC$.

Similarly, $NK \parallel CE$, U is the circumcenter of $\triangle KNT$, and $\angle UKT = 90° - \angle TNK = 90° - \angle BCD$.

Consequently, $\angle VKU = \angle VKT - \angle UKT = \angle BCD - \angle ABC$.

Solution 2   As shown in Fig. 6.3, extend EA to intersect the circumcircle of $\triangle BEK$ at P.

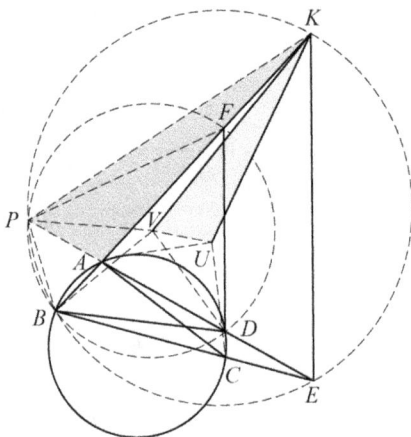

Fig. 6.3

By $\angle BPD = \angle BPE = \angle BKE = \angle BFD$, it follows that points P, B, D, F are concyclic.

By $\angle UAK = 90° - \angle ACD$ and $\angle VPB = 90° - \angle PDB$, it follows that

$$\angle KPV = \angle KPB - \angle VPB$$
$$= 180° - \angle KEB - \angle VPB$$
$$= 180° - \angle KEB - (90° - \angle PDB)$$
$$= 90° + \angle ACB - \angle DCB$$
$$= 90° - \angle ACD.$$

Therefore, $\angle UAK = \angle KPV$.

By $\angle BVD = 2\angle BFD = \angle AUC$, so isosceles $\triangle VBD \backsim$ isosceles $\triangle UAC$.

By $\triangle EBD \backsim \triangle EAC$, $\triangle VBD \backsim \triangle UAC$, so

$$\frac{PV}{AU} = \frac{BV}{AU} = \frac{BD}{AC} = \frac{BE}{AE} = \frac{PK}{AK}.$$

Thus, $\triangle KPV \backsim \triangle KAU$, and then $\angle PKV = \angle AKU$.
Therefore,

$$\angle VKU = \angle AKU - \angle AKV$$
$$= \angle PKV - \angle AKV = \angle PKA$$

$$= \angle AEB = \angle BCD - \angle CDE$$

$$= \angle BCD - \angle ABC.$$

Solution 3 As shown in Fig. 6.4, denote the center of the circle as O. Connect AU, FU, OA, OD, OE.

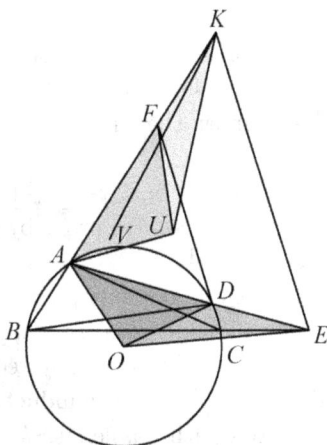

Fig. 6.4

Since $\angle AUF = 2\angle ACD = \angle AOD, UA = UF, OA = OD$, $\triangle AUF \backsim \triangle AOD$.

And since $CD \parallel EK$, we get $\frac{AD}{AE} = \frac{AF}{AK}$. Thus, $\triangle AUK \backsim \triangle AOE$. Therefore, $\angle AKU = \angle AEO$.

Similarly, there is $\angle AKV = \angle BEO$, so

$$\angle UKV = \angle UKA - \angle VKA = \angle OEA - \angle OEB = \angle AEB$$

$$= \angle BCD - \angle ABC.$$

④ Given the positive integer $k \geqslant l$. Let m be the smallest positive integer, satisfying: for any $n \geqslant m$, no matter how all edges of a complete graph of order n are colored in red and blue, there will always exist either a red path of length k or a blue path of length l. Prove that $m = k + \lfloor \frac{l+1}{2} \rfloor$.

Note: A path of length t consists of $t + 1$ points $U_1, U_2, \ldots, U_{t+1}$, satisfying that for any $1 \leqslant i \leqslant t$, there is an edge between U_i and U_{i+1}.

Solution On one hand, we give the construction of the coloring method when $n = k + \lfloor \frac{l+1}{2} \rfloor - 1$, satisfying that there does not exist a red path of length k or a blue path of length l. First select k points out of n points, and denote the set of these k points as S. Color the edges between any two of these k points red, and color all remaining edges blue. Since only k points are adjacent to the red edges, it is clear that the length of the red path does not exceed $k - 1$. Next, we prove that the length of the blue path does not exceed $l - 1$. For a blue path, note that adjacent points on the path cannot all be in S. And since this path passes through the points outside S at most $\lfloor \frac{l+1}{2} \rfloor - 1$ times, the length of this path does not exceed $2 \left(\lfloor \frac{l+1}{2} \rfloor - 1 \right) \leqslant l - 1$.

On the other hand, we show that $m \leqslant k + \lfloor \frac{l+1}{2} \rfloor$. Let $g(k, l)$ be the value of m corresponding to a general $k \geqslant l$. We prove by induction on k that for all $l \leqslant k$ there is $g(k, l) \leqslant k + \lfloor \frac{l+1}{2} \rfloor$.

When $k = 1$, the statement is obviously true. When $k \geqslant 2$, assume that the statement holds for all values less than k, and in the following we will prove that it also holds for k. Consider any red and blue 2-coloring a complete graph G of order $k + \lfloor \frac{l+1}{2} \rfloor$. By the inductive hypothesis, we need to prove that if the longest red path has length $k - 1$, then there exists a blue path of length l.

Suppose the longest red path is $U_1, U_2, \ldots, U_k$, and the remaining points are $V_1, V_2, \ldots, V_{\lfloor \frac{l+1}{2} \rfloor}$. Assume the sets $U = \{U_1, U_2, \ldots, U_k\}$, $V = \left\{ V_1, V_2, \ldots, V_{\lfloor \frac{l+1}{2} \rfloor} \right\}$. Then, by the maximality of the length of $U_1, U_2, \ldots, U_k$, it is easy to verify that the following three properties hold for any subscript within bounds:

(i) $V_i U_j$ is blue or $V_i U_{j+1}$ is blue;
(ii) $V_i U_j$ and $V_i U_{j+1}$ are both blue;
(iii) For j and three distinct i_1, i_2, i_3, at least one point in U_j, U_{j+1} is connected to at least two points in $V_{i_1}, V_{i_2}, V_{i_3}$ both by blue edges.

Consider the longest blue path satisfying the following properties: it does not contain U_1, U_k; both the starting point and the end point are in V; each pair of adjacent points has one point in U and the other point in V. Let this path be S, and the starting point and the end point be A and B, respectively. If S includes all points in V, then adding $U_1 A, B U_k$ to S will form a path of length $2 \lfloor \frac{l+1}{2} \rfloor \geqslant l$, which meets the requirements. Therefore, we might as well assume that S does not contain all points in V. Let W be the set of all points in V that are not in S.

Then we consider the longest blue path that satisfies the following properties: it does not contain U_1, U_k; it does not contain points in S; both the starting point and the end point are in W; each pair of adjacent points has one point in U and the other point in W. Let this path be T.

Next, we will prove that all points in V are either in S or T. Assume that there is a point X in V that is not in S or T. Then the number of points in S and T does not exceed $\lfloor \frac{l+1}{2} \rfloor - 1$. Therefore, the number of points of S and T in U does not exceed $\lfloor \frac{l+1}{2} \rfloor - 3 < \lfloor \frac{k-3}{2} \rfloor = \lfloor \frac{k-2-1}{2} \rfloor$ ($l \leqslant k$ is used). Therefore, there exists subscript $i, 2 \leqslant i \leqslant k-2$ such that U_i, U_{i+1} are not in S and T. Using property (iii) for $A, C, X \in V$ and $U_i, U_{i+1} \in U$, so we can extend S or T, contradicting the maximality of the path.

Therefore, all points in V are either in S or T, and the sum of the lengths of S and T is $2 \lfloor \frac{l+1}{2} \rfloor - 4$. Adding S and T to edges U_1A, BU_k, U_kC, DU_1 will form a cycle of length $2 \lfloor \frac{l+1}{2} \rfloor$. When l is odd, cut the cycle somewhere to get a cycle with a length of l. When l is even, note that the number of points of S and T in U does not exceed $\lfloor \frac{l+1}{2} \rfloor - 2 = \frac{l-4}{2} \leqslant \frac{k-4}{2}$. Therefore, there exists subscript $i, 2 \leqslant i \leqslant k-2$ such that U_i, U_{i+1} are not in S and T. Using property (i), we can get that U_i or U_{i+1} and the cycle just obtained are connected by blue edges. So by cutting the cycle we can get a cycle with a length of l. The proof is completed.

14:00–16:50; 28th July 2022

5 As shown in Fig. 6.5, in $\triangle ABC$, $AB = AC$, point D is on the extension of segment AB, and point E is on segment AC, satisfying $BD = CE$. Denote the circumcircle of $\triangle ABC$ as Γ. The circumcircle of $\triangle BDE$ intersects Γ at another point P, and the circumcircle of $\triangle CDE$ intersects Γ at another point Q. Prove that $PQ \parallel BC$.

Solution Take the midpoint M of minor arc $\overarc{BC}$ on circle Γ. Since $BM = CM, BD = CE, \angle DBM = \angle MCE$, we have $\triangle DBM \cong \triangle ECM$. So $\angle CEM = \angle MDA$, and hence points A, D, M, E are concyclic.

Therefore, the radical axes of the three circumcircles of $\triangle ADE, \triangle CQE$ and $\triangle ABC$, which are CQ, AM and DE, respectively, are concurrent.

Similarly, BP, AM and DE are concurrent. So CQ, BP, AM and DE are concurrent.

By symmetry, we have $PQ \parallel BC$.

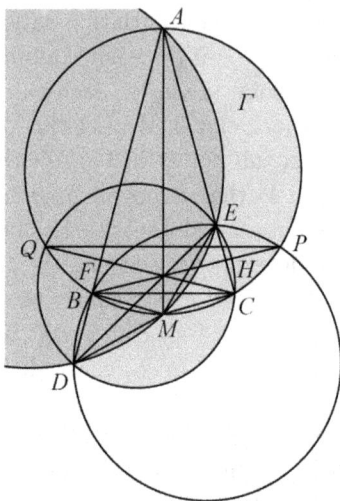

Fig. 6.5

6   Given $n \geqslant 2$ is a positive integer. Positive real numbers $b_1, b_2, \ldots, b_n$ satisfy $b_1 < b_2 < \cdots < b_n = 1$. Prove that

$$\frac{2}{3} + \sum_{k=1}^{n-1} \frac{b_k^2}{1 - b_{k+1} + b_k} < \sum_{k=1}^{n} b_k^2.$$

Solution

Lemma   When $0 < x < y \leqslant 1$, there is $\frac{x^2(y-x)}{1-y+x} \leqslant \frac{1}{3}(y^3 - x^3)$.

Proof of the lemma   We only need to prove $3x^2(y - x) \leqslant (y^3 - x^3)(1 - y + x)$, namely, $y^2 + xy - 2x^2 \geqslant y^3 - x^3$. That is, we only need to prove $y + 2x \geqslant x^2 + xy + y^2$.

Since $0 < x, y \leqslant 1$, we have $y \geqslant y^2$, $2x \geqslant x^2 + xy$, and hence the lemma is established.

Returning to the original problem, we only need to prove $\sum_{k=1}^{n-1} \frac{b_k^2}{1 - b_{k+1} + b_k} - \sum_{k=1}^{n-1} b_k^2 < \frac{1}{3}$.

Using the lemma, we know

$$\sum_{k=1}^{n-1} \frac{b_k^2}{1 - b_{k+1} + b_k} - \sum_{k=1}^{n-1} b_k^2 = \sum_{k=1}^{n-1} \frac{b_k^2 (b_{k+1} - b_k)}{1 - b_{k+1} + b_k} \leqslant \sum_{k=1}^{n-1} \frac{1}{3}(b_{k+1}^3 - b_k^3)$$

$$= \frac{1}{3} b_n^3 - \frac{1}{3} b_1^3 < \frac{1}{3}.$$

7 Prove that there exists a positive real number c such that for any positive integer n and n points on the plane, the number of isosceles triangles formed by these n points does not exceed $c \cdot n^{\frac{5}{2}}$.

Solution We first show that among the C_n^2 perpendicular bisectors of the n points, suppose there are k lines that coincide, and then there is $k \leqslant n$. In fact, if the perpendicular bisector of AB coincides with the perpendicular bisector of CD, then $\{A, B\} \cap \{C, D\} = \varnothing$ or $\{A, B\} = \{C, D\}$.

Consider these C_n^2 perpendicular bisectors. Suppose they are $l_1, l_2, \ldots, l_m$ after removing the duplicates, where the multiplicity of l_i is c_i, and then $c_i \leqslant n (1 \leqslant i \leqslant m)$. Suppose l_i passes through x_i points of the n points. Since two points determine a line, the point pairs (X, Y) on different l_i are different, so there is

$$C_{x_1}^2 + C_{x_2}^2 + \cdots + C_{x_m}^2 \leqslant C_n^2.$$

After simplifying, we get

$$\left(x_1 - \frac{1}{2}\right)^2 + \left(x_2 - \frac{1}{2}\right)^2 + \cdots + \left(x_m - \frac{1}{2}\right)^2 \leqslant n(n-1) + \frac{1}{4}m \leqslant 2n(n-1).$$

The number of isosceles triangles desired does not exceed $c_1 x_1 + c_2 x_2 + \cdots + c_m x_m$.

By the Cauchy inequality, we know

$$\left(c_1\left(x_1 - \frac{1}{2}\right) + c_2\left(x_2 - \frac{1}{2}\right) + \cdots + c_m\left(x_m - \frac{1}{2}\right)\right)^2$$

$$\leqslant (c_1^2 + c_2^2 + \cdots + c_m^2)\left(\left(x_1 - \frac{1}{2}\right)^2 + \left(x_2 - \frac{1}{2}\right)^2 + \cdots + \left(x_m - \frac{1}{2}\right)^2\right)$$

$$\leqslant 2n(n-1)(c_1^2 + c_2^2 + \cdots + c_m^2)$$

$$\leqslant 2n^2(n-1)(c_1 + c_2 + \cdots + c_m) = n^3(n-1)^2 < n^5.$$

Therefore, $c_1 x_1 + c_2 x_2 + \cdots + c_m x_m \leqslant n^{\frac{5}{2}} + \frac{1}{2}(c_1 + c_2 + \cdots + c_m) \leqslant 2n^{\frac{5}{2}}$.

8 Let $a_1, a_2, \ldots, a_{2022}$ be 2022 distinct prime numbers greater than 2022. Let $A_1, A_2, \ldots, A_{2022}$ be number sets and

$$A_i \subseteq \{1, 2, \ldots, a_i - 1\}(i = 1, 2, \ldots, 2022).$$

Prove that there exists a positive integer $m \leqslant (2|A_1| + 1)$ $(2|A_2| + 1) \cdots (2|A_{2022}| + 1)$ such that for any $i = 1, 2, \ldots, 2022$ and any $a \in A_i$, there is $m \not\equiv a \pmod{a_i}$.

Solution   Since $a_1, a_2, \ldots, a_{2022}$ are mutually coprime, it follows from the Chinese remainder theorem that for any $1, 2, \ldots, 2022$, there exists t_i that satisfies

$$t_i \equiv 1 \pmod{a_i}, \quad t_i \equiv 0 \pmod{a_j}(j \neq i).$$

Consider the number $\alpha = \sum_{i=1}^{2022} t_i x_i$, and then

$$\alpha \equiv x_i \pmod{a_i}. \qquad \text{①}$$

Denote $S_i = \{0, 1, 2, \ldots, a_i - 1\}$, and let $B_i \subseteq \{0, 1, 2, \ldots, a_i - 1\}$ such that for any $x, y \in B_i$ and any $z \in A_i$, $x - y \not\equiv z \pmod{a_i}$ is the set that has maximal number of elements. And for any $t \notin B_i, t \in S_i$, there must exist $u \in B_i$, $v \in A_i$ such that

$$\pm(t - u) \equiv v \pmod{a_i}.$$

Thus, $t \equiv u \pm v \pmod{a_i}$, so for any $t \notin B_i$, $t \in S_i$, t has at most $2|B_i||A_i|$ remainders modulo a_i.

Therefore, $|B_i| + 2|A_i||B_i| \geqslant a_i$, i.e., $|B_i| \geqslant \frac{a_i}{2|A_i|+1}, i = 1, 2, \ldots, 2022$.

Consider the set $S = \left\{ \sum_{i=1}^{2022} t_i x_i \mid x_i \in B_i \right\}$ composed of numbers of the form ①. Then, there are $|B_1||B_2| \cdots |B_{2022}|$ in S that are mutually incongruent modulo $a_1 a_2 \cdots a_{2022}$.

Denote the remainders of these numbers modulo $a_1 a_2 \cdots a_{2022}$ in ascending order as $0 \leqslant r_1 < r_2 < \cdots < r_{|B_1||B_2|\cdots|B_{2022}|} \leqslant a_1 a_2 \cdots a_{2022} - 1$.

Since $r_2 - r_1, r_3 - r_2, \ldots, r_{|B_1||B_2|\cdots|B_{2022}|} - r_{|B_1||B_2|\cdots|B_{2022}|-1}, r_1 - r_{|B_1||B_2|\cdots|B_{2022}|} + a_1 a_2 \cdots a_{2022}$ all belong to the set $0, a_1 a_2 \cdots a_{2022} - 1$ and their sum is $a_1 a_2 \cdots a_{2022}$, there must exist one of the differences, say m, satisfying

$$m \leqslant \frac{a_1 a_2 \cdots a_{2022}}{|B_1||B_2| \cdots |B_{2022}|} \leqslant (2|A_1| + 1)(2|A_2| + 1) \cdots (2|A_{2022}| + 1).$$

By the definition of B_i, we know that the difference of this number modulo a_i is not congruent to any element in A_i.

7

China National Team Selection Test

2023

The main task of the Chinese National Training Team for the 64th International Mathematical Olympiad (IMO) in 2023 was to select members of the Chinese national team for China to participate in the 64th International Mathematical Olympiad held in Chiba, Japan in 2023.

From 9th March to 18th March 2023, the first round selection was hosted by Chengdu Jiaxiang Foreign Languages Senior High School in Sichuan Province. The tests were conducted on 11th, 12th, 16th, and 17th, 8:00 am–12:30 pm on each day. A total of 60 contestants participated in the tests. After two tests, 19 contestants were selected to enter the second round.

The second round was held from 23rd March to 31st March 2023 at Hangzhou Xuejun High School, Wenyuan Campus in Zhejiang Province. During this period, two tests were conducted on 25th, 26th, 29th, and 30th, 8:00 am–12:30 pm on each day, and the total scores of the four tests were compared (with equal weights). The top six scorers were identified to be the Chinese national team members for the 64th IMO. The six members were Wang Chunji (Shanghai High School, 10th grade), Sun Qi'ao (Shanghai High School, 11th grade), Shi Haojia (Zhuji Hailiang Senior High School, 10th grade), Zhang Xinliang (Zhenhai High School of Ningbo, 12th grade), Jiang Zhicheng (Shenzhen Middle School, 11th grade), and Liang Xingjian (The High School Attached to Hunan Normal University, 12th grade).

The coaches of the national training team are (in alphabet order): Ai Yinghua (Tsinghua University), Fu Yunhao (Southern University of

Science and Technology), He Yijie (East China Normal University), Li Ting (Sichuan University), Qu Zhenhua (East China Normal University), Wang Bin (Academy of Mathematics and Systems Science, Chinese Academy of Sciences), Wang Xinmao (University of Science and Technology of China), Xiao Liang (Peking University), Xiong Bin (East China Normal University), Yao Yijun (Fudan University), Yu Hongbing (Soochow University), and Zhang Sihui (University of Shanghai for Science and Technology).

Test I, First Day
(8 am–12:30 pm; 11 March 2023)

1   Given an integer $n \geqslant 2$, suppose there a point P inside a convex cyclic $2n$-gon $A_1 A_2 \cdots A_{2n}$ such that

$$\angle P A_1 A_2 = \angle P A_2 A_3 = \cdots = \angle P A_{2n-1} A_{2n} = \angle P A_{2n} A_1.$$

Prove that

$$\prod_{i=1}^{n} |A_{2i-1} A_{2i}| = \prod_{i=1}^{n} |A_{2i} A_{2i+1}|,$$

where $A_{2n+1} = A_1$.

(Contributed by Xiao Liang)

Solution 1   As shown in Fig. 7.1, all subscripts are modulo $2n$. Denote $\angle P A_1 A_2 = \angle P A_2 A_3 = \cdots = \alpha$. For $1 \leqslant i \leqslant 2n$, extend $A_i P$ to intersect the circle at point B_i. Note that

$$\angle A_i B_i B_{i-1} = \angle A_i A_{i-1} P = \angle A_{i+1} A_i B_i = \alpha.$$

Hence, $A_i A_{i+1} B_{i-1} B_i$ is an isosceles trapezium. (In fact, $B_1 B_2 \cdots B_{2n}$ is obtained by rotating $A_2 A_3 \cdots A_{2n} A_1$ along the circle clockwise by 2α.)

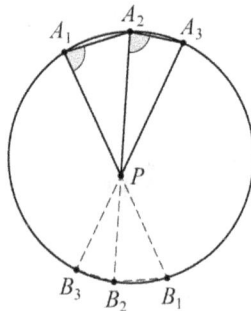

Fig. 7.1

In particular, we have $A_i A_{i+1} = B_{i-1} B_i$. Moreover, as $\triangle A_{i-1} P A_i$ and $\triangle B_i P B_{i-1}$ are similar, it follows that

$$\frac{A_{i-1}P}{B_i P} = \frac{A_i P}{B_{i-1}P} = \frac{A_{i-1}A_i}{B_{i-1}B_i} = \frac{A_{i-1}A_i}{A_i A_{i+1}}, \quad 1 \leqslant i \leqslant 2n. \qquad \textcircled{1}$$

Taking all odd i's, the product of the fourth fractions in $\textcircled{1}$ equals the product of the first fractions, which gives

$$\prod_{i=1}^{n} \frac{A_{2i-2}A_{2i-1}}{A_{2i-1}A_{2i}} = \prod_{i=1}^{n} \frac{A_{2i-2}P}{B_{2i-1}P}.$$

Taking all even i's, the product of the fourth fractions in $\textcircled{1}$ equals the product of the second fractions, which gives

$$\prod_{i=1}^{n} \frac{A_{2i-1}A_{2i}}{A_{2i}A_{2i+1}} = \prod_{i=1}^{n} \frac{A_{2i}P}{B_{2i-1}P}.$$

Note that in the above two equations, the right-hand sides are equal. Therefore,

$$\prod_{i=1}^{n} \frac{A_{2i-2}A_{2i-1}}{A_{2i-1}A_{2i}} = \prod_{i=1}^{n} \frac{A_{2i-1}A_{2i}}{A_{2i}A_{2i+1}} \Rightarrow \prod_{i=1}^{n} (A_{2i}A_{2i+1})^2 = \prod_{i=1}^{n} (A_{2i-1}A_{2i})^2,$$

yielding the desired identity.

Solution 2 As illustrated in Fig. 7.2, let H be the intersection point of $A_1 A_4$ and line $P A_3$. Since A_1, A_2, A_3 and A_4 are concyclic,

$$\angle A_2 A_1 A_4 = 180° - \angle A_2 A_3 A_4$$
$$= 180° - \angle A_2 A_3 P - \angle P A_3 A_4$$
$$= 180° - \angle A_2 A_3 P - \angle A_3 A_2 P$$
$$= \angle A_2 P A_3.$$

It follows that A_1, A_2, H and P are concyclic, in particular, $\angle A_2 H A_3 = \angle A_2 A_1 P = \angle H A_3 A_4$, and thus $A_2 H \parallel A_3 A_4$. As A_1, A_2, H and P are concyclic, we have

$$\angle A_3 A_4 H = \angle A_2 H A_1 = \angle A_1 P A_2.$$

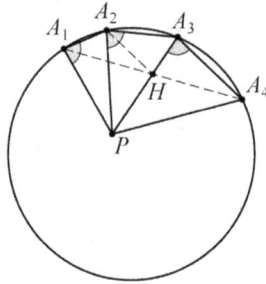

Fig. 7.2

Together with $\angle A_2 A_1 P = \angle A_4 A_3 H$, it follows that $\triangle A_1 A_2 P$ and $\triangle A_3 H A_4$ are similar, and

$$\frac{A_1 A_2}{A_1 P} = \frac{A_3 H}{A_3 A_4} \Rightarrow A_1 A_2 \cdot A_3 A_4 = A_1 P \cdot A_3 H.$$

On the other hand, from $\angle A_3 H A_2 = \angle A_2 A_1 P = \angle A_3 A_2 P$, we see that $\triangle A_3 H A_2$ and $\triangle A_3 A_2 P$ are similar, giving

$$\frac{A_2 A_3}{A_3 H} = \frac{P A_3}{A_2 A_3} \Rightarrow (A_2 A_3)^2 = P A_3 \cdot A_3 H.$$

Taking the quotient of the above two equations, we arrive at

$$\frac{A_1 A_2 \cdot A_3 A_4}{(A_2 A_3)^2} = \frac{P A_1}{P A_3}.$$

The same argument can be applied to $A_{2i-1}, A_{2i}, A_{2i+1}$ and A_{2i+2} to obtain similar relations as above. Multiplying them to reach

$$\prod_{i=1}^{n} \frac{A_{2i-1} A_{2i} \cdot A_{2i+1} A_{2i+2}}{(A_{2i} A_{2i+1})^2} = \prod_{i=1}^{n} \frac{P A_{2i-1}}{P A_{2i+1}} = 1,$$

where the subscripts are modulo $2n$. Now it can be reduced to

$$\prod_{i=1}^{n} |A_{2i-1} A_{2i}|^2 = \prod_{i=1}^{n} |A_{2i} A_{2i+1}|^2,$$

as desired.

Solution 3   Suppose that the circumscribed circle is the unit circle in the complex plane, and the points $A_1, \ldots, A_{2n}$ correspond to complex numbers $z_1, \ldots, z_{2n}$, respectively. Without loss of generality (apply a rotation if necessary), assume that point P corresponds to a nonzero real number $a < 1$, and denote $\angle P A_1 A_2 = \pi - \alpha \in (0, \pi)$.

By the given condition, $\frac{z_2-z_1}{(z_1-a)e^{i\alpha}} \in \mathbf{R}$, and thus

$$\frac{z_2 - z_1}{(z_1 - a)e^{i\alpha}} = \frac{\overline{z_2} - \overline{z_1}}{(\overline{z_1} - a)e^{-i\alpha}}.$$

As z_1, z_2 are unit complex numbers, $z_1 \neq z_2$, we can find $z_2 = \frac{e^{2i\alpha}(z_1-a)}{az_1-1}$, and likewise $z_{j+1} = \frac{e^{2i\alpha}(z_j-a)}{az_j-1}$ $(j = 1, 2, \ldots, 2n)$, in which the subscripts are modulo $2n$. If $a = 0$, then $z_{j+1} = -e^{2i\alpha}z_j$, $A_1A_2\cdots A_{2n}$ is a regular $2n$-gon, and the conclusion is obvious. In the following, assume $0 < a < 1$.

Consider the fractional linear transformation $f(z) = \frac{e^{2i\alpha}(z-a)}{az-1}$, which has two fixed points (as the two roots of the quadratic equation $ax^2 - (1 + e^{2i\alpha})x + ae^{2i\alpha} = 0$) say x_1 and x_2. It is easy to verify that $x_1 \neq x_2$.

Let $\frac{z_j-x_1}{z_j-x_2} = w_j$. As $f(z_j) = z_{j+1}$, we see that $w_1, w_2, \ldots, w_{2n}, w_1$ form a geometric sequence, that is, the common ratio is a constant. Now, from

$$z_j = \frac{x_1 - x_2 w_j}{1 - w_j} = x_2 \cdot \frac{w_j - \frac{x_1}{x_2}}{w_j - 1},$$

we have

$$z_{j+1} - z_j = \frac{x_2}{(w_j - 1)(w_{j+1} - 1)}$$

$$\times \left(\left(w_{j+1} - \frac{x_1}{x_2}\right)(w_j - 1) - \left(w_j - \frac{x_1}{x_2}\right)(w_{j+1} - 1)\right)$$

$$= \frac{(x_1 - x_2)(w_{j+1} - w_j)}{(w_j - 1)(w_{j+1} - 1)}.$$

Consequently,

$$\prod_{j=1}^{n} \frac{z_{2j+1} - z_{2j}}{z_{2j} - z_{2j-1}} = \prod_{j=1}^{n} \frac{w_{2j+1} - w_{2j}}{w_{2j} - w_{2j-1}} = \prod_{j=1}^{n} \frac{w_{2j}}{w_{2j-1}}$$

is a unit complex number (in fact, it is -1), and the conclusion follows.

② N people attend a party. There are at most n pairs of friends among them, and every two people shake hands if and only if they have at least one common friend.

Given a positive integer $m \geqslant 3$ such that $m^3 \geqslant n$. Show that there exists a person A, the number of people that shake hands with is no more than $m - 1$ times the number of A's friends.

(Contributed by Wang Bin)

Solution We introduce a graph that each vertex represents a person and each edge connects a pair of friends.

Clearly, there exists a connected component $G = (V, E)$ satisfying $|E| \leqslant |V| \leqslant n$. In G, for each vertex v, let $d(v)$ be the degree of v and $\beta(v)$ be the number of persons shaking hands with v.

Suppose the statement does not hold. For every vertex v, we have

$$\beta(v) \geqslant (m-1) \cdot d(v) + 1. \qquad ②$$

Here is the key observation: for every vertex v_0 of degree 1, say v_0 is adjacent to v_1, ② indicates that

$$\beta(v_0) \geqslant (m-1) \cdot d(v_0) + 1 = m,$$

and thus $d(v_1) \geqslant m + 1$.

Remove all degree-1 vertices and edges incident to them to obtain $G' = (V', E')$. Note that the numbers of the removed vertices and edges are equal, so G' is connected and $|E'| \leqslant |V'|$, implying that G' is a tree or has a unique cycle. There are three situations.

(1) G' is a single vertex v_1. By the key observation, $d(v_1) \geqslant m + 1$. Then G includes $d(v_1) + 1$ vertices and every edge of G is incident to v_1. We have

$$\beta(v_1) = 0 \leqslant (m-1)d(v_1),$$

contradicting the assumption at the beginning of the proof.

(2) G' includes a vertex v_1 of degree 1. Then in G, v_1 is adjacent to some vertex v_0 of degree 1. By the key observation, $d(v_1) \geqslant m + 1$, that is to say in G, v_1 is adjacent to $d(v_1) - 1 \geqslant m$ vertices of degree 1. Suppose $v_1 v_2$ is an edge in G'. The inequality ② for v_1 as

$$\beta(v_1) \geqslant (m-1)d(v_1) + 1$$

becomes

$$d(v_2) - 1 \geqslant (m-1)d(v_1) + 1 \geqslant m^2,$$

i.e., $d(v_2) \geqslant m^2 + 1$.

We further apply the inequality ② to v_2 to yield

$$\beta(v_2) \geqslant (m-1)d(v_2) + 1 \geqslant (m-1)(m^2 + 1) + 1.$$

Note that the friends of v_2 and the persons who shake hands with v_2 are all distinct, unless v_2 lies on a cycle of length 3 for which v_2 shakes hands with his/her two friends, which is impossible. This indicates that G includes at least

$$1 + d(v_2) + \beta(v_2) - 2 \geqslant 1 + (m^2 + 1) + (m - 1)(m^2 + 1) + 1 - 2$$
$$= m^3 + m > m^3$$

vertices, contradicting the condition $n \leqslant m^3$.

(3) The degree of every vertex of G' is at least 2. Then G' is a cycle $v_1 \sim v_2 \sim \cdots \sim v_t \sim v_1$, $t > 2$. Take the subscripts modulo t. By the key observation, every vertex v_i is adjacent to exactly $d(v_i) - 2$ vertices of degree 1 in G. Now ② applied to v_i gives

$$(d(v_{i-1}) - 2) + (d(v_{i+1}) - 2) + 2 \geqslant \beta(v_i)$$
$$\geqslant (m - 1) \cdot d(v_i) + 1 \geqslant 2d(v_i) + 1.$$

Summing the above inequalities over $i = 1, 2, \ldots, t$ will lead to a contradiction.

To sum up, we have shown that some vertex v must satisfy $\beta(v) \leqslant (m - 1)d(v)$, as desired.

3 (1) Given coprime positive integers a and b. Prove that there exist real numbers λ and β, such that for any positive integer m,

$$\lambda m - \beta \leqslant \sum_{k=1}^{m-1} \left\{ \frac{ak}{m} \right\} \cdot \left\{ \frac{bk}{m} \right\} \leqslant \lambda m + \beta. \qquad ③$$

(2) Prove that there exists a positive integer N, such that for any prime number $p > N$, the following proposition holds.

Proposition *If positive integers a, b and c satisfy that $(a+b)(a+c)(b+c)$ is not divisible by p, then in the set $\{1, 2, \ldots, p-1\}$, there are at least $\lfloor \frac{p}{12} \rfloor$ elements k such that*

$$\left\{ \frac{ak}{p} \right\} + \left\{ \frac{bk}{p} \right\} + \left\{ \frac{ck}{p} \right\} \leqslant 1.$$

Here, $\lfloor x \rfloor$ is the largest integer not exceeding x, and $\{x\} = x - \lfloor x \rfloor$ is the decimal part of x.

(Contributed by Wang Bin)

Solution   (1) When m is large, we expect the summation in ③ to be very close to the following integral:

$$\int_0^m \left\{\frac{ax}{m}\right\} \left\{\frac{bx}{m}\right\} \mathrm{d}x = m \cdot \int_0^1 \{ax\}\{bx\}\mathrm{d}x.$$

Henceforth we guess and prove that ③ holds for

$$\lambda = \int_0^1 \{ax\}\{bx\}\mathrm{d}x, \quad \beta = 2(a+b).$$

First, we estimate the difference

$$\left| \sum_{k=1}^{m-1} \left\{\frac{ak}{m}\right\} \left\{\frac{bk}{m}\right\} - \int_0^m \left\{\frac{ax}{m}\right\} \left\{\frac{bx}{m}\right\} \mathrm{d}x \right|$$

$$= \left| \sum_{k=0}^{m-1} \frac{1}{2} \left(\left\{\frac{ak}{m}\right\} \left\{\frac{bk}{m}\right\} + \left\{\frac{a(k+1)}{m}\right\} \left\{\frac{b(k+1)}{m}\right\} \right) \right.$$

$$\left. - \int_0^m \left\{\frac{ax}{m}\right\} \left\{\frac{bx}{m}\right\} \mathrm{d}x \right|$$

$$\leqslant \sum_{k=0}^{m-1} \left| \frac{1}{2} \left(\left\{\frac{ak}{m}\right\} \left\{\frac{bk}{m}\right\} + \left\{\frac{a(k+1)}{m}\right\} \left\{\frac{b(k+1)}{m}\right\} \right) \right.$$

$$\left. - \int_k^{k+1} \left\{\frac{ax}{m}\right\} \left\{\frac{bx}{m}\right\} \mathrm{d}x \right|. \qquad ④$$

It suffices to show this bound is less than or equal to $2(a+b)$.

If $m \leqslant a+b$, every term in ④ does not exceed 1, and therefore the sum is less than or equal to $2a + 2b$. In the following, we assume $m > a + b$.

For each k, there are two possible situations:

(a) If for $x \in (k, k+1]$, the decimal function $\left\{\frac{ax}{m}\right\}$ or $\left\{\frac{bx}{m}\right\}$ is not continuous. There are at most $a+b-1$ of such k's. Also, as the product $\left\{\frac{ax}{m}\right\} \left\{\frac{bx}{m}\right\}$ always takes values in $[0, 1]$, the corresponding terms will contribute no more than $a+b-1$ to the bound in ④.

(b) If for $x \in (k, k+1]$, $\left\{\frac{ax}{m}\right\}$ or $\left\{\frac{bx}{m}\right\}$ is continuous. Let

$$x = k + \delta, \quad \delta \in (0, 1].$$

Then $\left\{\frac{ax}{m}\right\} = \left\{\frac{ak}{m}\right\} + \frac{a\delta}{m}$ and $\left\{\frac{bx}{m}\right\} = \left\{\frac{bk}{m}\right\} + \frac{b\delta}{m}$. Denote $\alpha = \left\{\frac{ak}{m}\right\}$, $\beta = \left\{\frac{bk}{m}\right\}$. The corresponding term in ④ is

$$\frac{1}{2}\left(\left\{\frac{ak}{m}\right\}\left\{\frac{bk}{m}\right\} + \left\{\frac{a(k+1)}{m}\right\}\left\{\frac{b(k+1)}{m}\right\}\right)$$

$$- \int_k^{k+1} \left\{\frac{ax}{m}\right\}\left\{\frac{bx}{m}\right\} \mathrm{d}x$$

$$= \frac{1}{2}\left(\alpha\beta + \left(\alpha + \frac{a}{m}\right)\left(\beta + \frac{b}{m}\right)\right) - \int_0^1 \left(\alpha + \frac{a\delta}{m}\right)\left(\beta + \frac{b\delta}{m}\right)\mathrm{d}\delta$$

$$= \left(\alpha\beta + \frac{\alpha\beta}{2m} + \frac{b\alpha}{2m} + \frac{ab}{2m^2}\right) - \left(\alpha\beta + \frac{\alpha\beta}{2m} + \frac{b\alpha}{2m} + \frac{ab}{3m^2}\right) = \frac{ab}{6m^2}.$$

Combining the results in (a) and (b), we infer that the bound in ④ (when $m > a + b$) does not exceed

$$a + b - 1 + m \cdot \frac{ab}{6m^2} < 2(a+b) - 1.$$

This verifies ③.

(2) **Step 1:** A sharper estimate of the summation in Solution 1.

We first evaluate $\lambda = \int_0^1 \{ax\}\{bx\}\mathrm{d}x$. Intuitively, this is equivalent to integrating along the diagonal of an $a \times b$ grid, for example, $a = 4, b = 3$ as shown in the following left figure. Since the integrand only depends on the unit square that holds the current portion of the diagonal, by translation (or taking modulo 1) we can move the diagonal entirely in one unit square, as in the middle figure (the size of the unit square is enlarged). But this square can be further divided into ab small rectangles such that the integral path is along the diagonals, as illustrated in the following right figure.

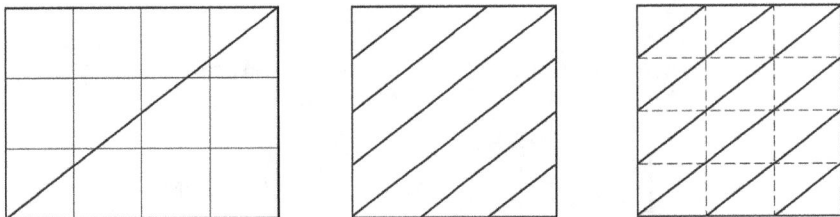

Essentially, the grid divides the integral from 0 to 1 into integrals from $\frac{i}{ab}$ to $\frac{i+1}{ab}(0 \leqslant i \leqslant ab - 1)$. Note that $\{ax\}$ and $\{bx\}$ are continuous in each unit square. So we have

$$\lambda = \int_0^1 \{ax\}\{bx\}\mathrm{d}x = \frac{1}{ab} \sum_{i=0}^{ab-1} \int_i^{i+1} \left\{\frac{x}{b}\right\}\left\{\frac{x}{a}\right\} \mathrm{d}x$$

$$= \frac{1}{ab} \sum_{i=0}^{ab-1} \left\{\frac{i+\frac12}{b}\right\}\left\{\frac{i+\frac12}{a}\right\}$$

$$+ \frac{1}{ab} \sum_{i=0}^{ab-1} \int_i^{i+1} \left(\left\{\frac{x}{b}\right\}\left\{\frac{x}{a}\right\} - \left\{\frac{i+\frac12}{b}\right\}\left\{\frac{i+\frac12}{a}\right\}\right) \mathrm{d}x, \qquad \text{⑤}$$

where in the second summation each integral can be evaluated as

$$\int_{-\frac12}^{\frac12} \left(\left\{\frac{i+\frac12+x}{b}\right\}\left\{\frac{i+\frac12+x}{a}\right\} - \left\{\frac{i+\frac12}{b}\right\}\left\{\frac{i+\frac12}{a}\right\}\right) \mathrm{d}x$$

$$= \int_{-\frac12}^{\frac12} \left(\left\{\frac{i+\frac12}{b}\right\}\frac{x}{a} + \left\{\frac{i+\frac12}{a}\right\}\frac{x}{b} + \frac{x^2}{ab}\right) \mathrm{d}x$$

$$= 0 + 0 + \frac{1}{3ab}\left(\left(\frac12\right)^3 - \left(-\frac12\right)^3\right) = \frac{1}{12ab}.$$

For the first summation in ⑤, note that as i traverses $\{0, 1, \ldots, ab - 1\}$, $(\{\frac{i}{b}\}, \{\frac{i}{a}\})$ takes every pair of

$$\left\{0, \frac{1}{a}, \ldots, \frac{a-1}{a}\right\} \times \left\{0, \frac{1}{b}, \ldots, \frac{b-1}{b}\right\}.$$

Therefore,

$$\sum_{i=0}^{ab-1} \left\{\frac{i+\frac12}{b}\right\}\left\{\frac{i+\frac12}{a}\right\} = \sum_{j=0}^{a-1}\sum_{k=0}^{b-1} \left\{\frac{k+\frac12}{b}\right\}\left\{\frac{j+\frac12}{a}\right\} = \frac{a}{2} \cdot \frac{b}{2} = \frac{ab}{4}.$$

Now we sum up the results and obtain

$$\lambda = \frac{1}{ab} \cdot \frac{ab}{4} + \frac{1}{ab} \cdot ab \cdot \frac{1}{12ab} = \frac14 + \frac{1}{12ab},$$

which gives

$$\left|\sum_{k=1}^{m-1} \left\{\frac{ak}{m}\right\}\left\{\frac{bk}{m}\right\} - \left(\frac{m}{4} + \frac{m}{12ab}\right)\right| \leqslant 2|a| + 2|b| - 1.$$

When m is coprime with both a and b, as $\{\frac{ak}{m}\} + \{\frac{-ak}{m}\} = 1$, the two summations corresponding to (a, b) and $(-a, b)$ have sum equal to $\frac{m-1}{2}$, and thus for a, b coprime with m (possibly negative), we have

$$\left| \sum_{k=1}^{m-1} \left\{ \frac{ak}{m} \right\} \left\{ \frac{bk}{m} \right\} - \left(\frac{m}{4} + \frac{m}{12ab} \right) \right| \leqslant 2|a| + 2|b|. \tag{$*$}$$

Step 2: Convert the problem into a summation similar to Solution 1.

Assume that the prime number p is sufficiently large, for example, $p > 10^{15}$. The symbol $\equiv$ in the context is taken modulo p.

Rename a, b and c as a_1, a_2 and a_3, respectively. Take a_4 with $a_1 + a_2 + a_3 + a_4 \equiv 0$. For $k \in \{1, 2, \ldots, p-1\}$, define

$$h_k = \left\{ \frac{a_1 k}{p} \right\} + \left\{ \frac{a_2 k}{p} \right\} + \left\{ \frac{a_3 k}{p} \right\} + \left\{ \frac{a_4 k}{p} \right\}.$$

Evidently, h_k is an integer, $0 \leqslant h_k < 4$, $h_k \in \{0, 1, 2, 3\}$. Also, $\left\{ \frac{a_1 k}{p} \right\} + \left\{ \frac{a_2 k}{p} \right\} + \left\{ \frac{a_3 k}{p} \right\} \leqslant 1$ if and only if $h_k = 0$ or 1.

Suppose that among $h_1, h_2, \ldots, h_{p-1}$ there are exactly L_0 of 0's, L_1 of 1's, L_2 of 2's and L_3 of 3's. It is required to show $L_0 + L_1 \geqslant \lfloor \frac{p}{12} \rfloor$.

If a_1, a_2, a_3 and a_4 are all multiples of p, then $L_0 = p - 1$ and the conclusion follows; if exactly two of them are multiples of p, then $L_1 = p-1$ and the conclusion follows; if exactly one of them is a multiple of p, then $h_{p-k} = 3 - h_k$, $L_1 = L_2 = \frac{p-1}{2}$, and the conclusion follows as well.

Now we assume $a_1, a_2, a_3, a_4 \not\equiv 0$. Then $h_{p-k} = 4 - h_k$, and

$$L_0 = 0, \quad L_1 = L_3, \quad L_1 + L_2 + L_3 = p - 1.$$

Treat $\left\{ \frac{ak}{p} \right\}$ as a random variable. Loosely speaking, the estimate of L_1 is related to the variance of summing h_k. Specifically, we consider the sum of squares

$$H = \sum_{k=1}^{p-1} h_k^2 = L_1 + 4L_2 + 9L_3 = 4(p-1) + 2L_1.$$

On the other hand, we have

$$H = \sum_{k=1}^{p-1} h_k^2 = \sum_{k=1}^{p-1} \left(\sum_{i=1}^{4} \left\{ \frac{a_i k}{p} \right\} \right) \left(\sum_{j=1}^{4} \left\{ \frac{a_j k}{p} \right\} \right)$$

$$= \sum_{i=1}^{4} \sum_{j=1}^{4} \left(\sum_{k=1}^{p-1} \left\{ \frac{a_i k}{p} \right\} \left\{ \frac{a_j k}{p} \right\} \right).$$

For any $a, b \not\equiv 0$, define function $f(a, b) = \sum_{k=1}^{p-1} \left\{ \frac{ak}{p} \right\} \left\{ \frac{bk}{p} \right\}$. In particular,

$$f(a_i, a_i) = f(1, 1) = \frac{1^2 + 2^2 + \cdots + (p-1)^2}{p^2} = \frac{p}{3} - \frac{1}{2} + \frac{1}{6p}.$$

Now consider the six-term sum: $F = \sum_{1 \leqslant i < j \leqslant 4} f(a_i, a_j)$. Since $H = 2F + 4f(1, 1) = 4(p-1) + 2L_1$, our goal $L_1 \geqslant \lfloor \frac{p}{12} \rfloor$ is achieved once we can verify $F \geqslant \frac{17}{12} p$.

Step 3: Turn to estimate a_1, a_2, a_3 and a_4.

By definition of f and the inequality $(*)$ in step 1 (taking $m = p$), we have

$$\left| f(a_i, a_j) - \left(\frac{p}{4} + \frac{p}{12 a_i a_j} \right) \right| \leqslant 2(|a_i| + |a_j|). \qquad (**)$$

Note when $|a_i|, |a_j|$ are large, the estimate of $f(a_i, a_j)$ is not very good.

We define an equivalence relation between primitive remainder pairs modulo p: $(a, b) \sim (u, v)$ if there exists $r \not\equiv 0$ such that $u \equiv ra, v \equiv rb$. Then,

$$f(u, v) = \sum_{k=1}^{p-1} \left\{ \frac{uk}{p} \right\} \left\{ \frac{vk}{p} \right\} = \sum_{k=1}^{p-1} \left\{ \frac{ark}{p} \right\} \left\{ \frac{brk}{p} \right\}$$

$$= \sum_{rk \equiv 1} \left\{ \frac{a \cdot rk}{p} \right\} \left\{ \frac{b \cdot rk}{p} \right\} = f(a, b),$$

which indicates that the value of $f(a_i, a_j)$ only depends on the equivalence class of (a_i, a_j). Returning to the estimate $(**)$, we wish to find another pair (u, v) in the class of (a_i, a_j) that has smaller absolute values.

For any $a, b \not\equiv 0$, denote b^{-1} as the reciprocal of b modulo p (so $bb^{-1} \equiv 1$). Among the remainders of $0, ab^{-1}, 2 \times ab^{-1}, \ldots, \lfloor \sqrt{p} \rfloor \times ab^{-1}$ (mod p), according to the pigeonhole principle there exist two of them whose difference is $\leqslant \lfloor \sqrt{p} \rfloor$, and we have

$$u, v \in \{\pm 1, \pm 2, \ldots, \pm \lfloor \sqrt{p} \rfloor\}$$

such that $u \equiv v \times ab^{-1}$, that is, $(u, v) \sim (a, b)$ (assume $u > 0$ and u, v coprime; otherwise replace by $\frac{u}{\gcd(u,v)}, \frac{v}{\gcd(u,v)}$). For (a_i, a_j), let (u_{ij}, v_{ij}) be the (u, v) pair found in the above process; if not unique, any pair will suffice.

We have $f(a_i, a_j) = f(u_{ij}, v_{ij})$, and by (**),

$$\left| f(a_i, a_j) - \left(\frac{p}{4} + \frac{p}{12 u_{ij} v_{ij}} \right) \right| < 4\sqrt{p}.$$

Let $g_{i,j} = \frac{1}{u_{ij} v_{ij}}$ and

$$G = \sum_{1 \leqslant i < j \leqslant 4} g_{i,j} = \sum_{1 \leqslant i < j \leqslant 4} \frac{1}{u_{ij} v_{ij}}.$$

Take $\delta = 10^{-5}$. If we can find $g_{i,j}$ (by some suitable choice of $(u_{ij}, v_{ij}) \sim (a_i, a_j)$) such that $G \geqslant -1 + \delta$, then

$$F = \sum_{i<j} f(a_i, a_j) > \sum_{i<j} \left(\frac{p}{4} + \frac{p}{12 u_{ij} v_{ij}} - 4\sqrt{p} \right)$$

$$= \frac{18 + G}{12} \times p - 24\sqrt{p} > \frac{17}{12} p.$$

The problem condition $p \nmid (a+b)(a+c)(b+c)$ guarantees that for each pair (a_i, a_j), as $a_i + a_j \not\equiv 0$, $(u_{ij}, v_{ij}) \neq (1, -1)$, it is always true that $g_{i,j} \geqslant -\frac{1}{2}$. Call (a, b) a *nice pair*, if there exist $u, v \in \{\pm 1, \pm 2, \ldots, \pm 10\}$ such that $(a, b) \sim (u, v)$. Clearly, if (a_i, a_j) is not a nice pair, then $g_{i,j} = \frac{1}{u_{ij} v_{ij}} \geqslant -\frac{1}{11}$. This indicates that among six pairs $(a_i, a_j)(1 \leqslant i < j \leqslant 4)$, if there is 0 or 1 nice pair, then $G \geqslant -\frac{1}{2} + 5 \times \left(-\frac{1}{11} \right) = -\frac{21}{22} > -1 + \delta$ and the proof is done.

Suppose there are two or more nice pairs among six pairs (a_i, a_j). From

$$a_1 + a_2 + a_3 + a_4 \equiv 0,$$

we see that the four numbers are proportional to simple expressions of u, v. For instance,

- If $(a_1, a_2) \sim (u_1, v_1), (a_1, a_3) \sim (u_2, v_2)$, then they are proportional to $(u_1 u_2, v_1 u_2, v_2 u_1, -u_1 u_2 - v_1 u_2 - v_2 u_1)$;
- If $(a_1, a_2) \sim (u_1, v_1), (a_3, a_4) \sim (u_2, v_2)$, then they are proportional to $(u_1(u_2 + v_2), v_1(u_2 + v_2), -u_2(u_1 + v_1), -v_2(u_1 + v_1))$.

Anyhow, there exist $z_1 + z_2 + z_3 + z_4 = 0$ satisfying

$$(a_1, a_2, a_3, a_4) \sim (z_1, z_2, z_3, z_4)$$

and $1 \leqslant |z_i| \leqslant 300$, $z_i + z_j \neq 0$.

Now we have $(a_i, a_j) \sim \left(\frac{z_i}{\gcd(z_i, z_j)}, \frac{z_j}{\gcd(z_i, z_j)} \right)$, and this natural choice gives $g_{i,j} = \frac{\gcd(z_i, z_j)^2}{z_i z_j}$. We are still concerned with the value of $G = \sum_{i<j} g_{i,j}$: the simple proportions that lead to $G = -1$ are of two types:

- The four numbers are proportional to *the first golden ratio* $(1 : -2 : -3 : 4)$. Let $a_1 = p - 1, a_2 = 2, a_3 = 3$ and $a_4 = p - 4$. Then $h_k = 1$ if and only if $\left\{\frac{2k}{p}\right\} + \left\{\frac{3k}{p}\right\} \leqslant \left\{\frac{k}{p}\right\}$. In $\{1, 2, \ldots, p-1\}$, these k's exactly satisfy $\frac{2}{3}p < k < \frac{3}{4}p$, and there are

$$L_1 = \left\lfloor \frac{3}{4}p \right\rfloor - \left\lfloor \frac{2}{3}p \right\rfloor \geqslant \left\lfloor \frac{p}{12} \right\rfloor$$

of them.

- The four numbers are proportional to *the second golden ratio* $(1 : -3 : -4 : 6)$. Let $a_1 = p - 1, a_2 = 3, a_3 = 4, a_4 = p - 6$. Then $h_k = 1$ if and only if $\left\{\frac{3k}{p}\right\} + \left\{\frac{4k}{p}\right\} \leqslant \left\{\frac{k}{p}\right\}$. In $\{1, 2, \ldots, p-1\}$, these k's exactly satisfy $\frac{3}{4}p < k < \frac{5}{6}p$, and there are

$$L_1 = \left\lfloor \frac{5}{6}p \right\rfloor - \left\lfloor \frac{3}{4}p \right\rfloor \geqslant \left\lfloor \frac{p}{12} \right\rfloor$$

of them.

We claim if $(a_1, a_2, a_3, a_4) \sim (z_1, z_2, z_3, z_4)$ is not one of the golden ratios described as above, then $G \geqslant -1 + \delta$. For the signs of z_1, z_2, z_3 and z_4,

(i) If three are positive and one is negative or vice versa, assume $z_1, z_2, z_3 > 0, z_4 = -(z_1 + z_2 + z_3)$. Then $G = \sum_{1 \leqslant i < j \leqslant 4} g_{i,j}$ consists of three positive and three negative terms, in which

$$g_{i,4} = -\frac{\gcd(z_i, z_4)^2}{z_i|z_4|} \geqslant -\frac{z_i}{|z_4|}, \quad i = 1, 2, 3,$$

and $g_{1,2} \geqslant \frac{1}{z_1 z_2} > \delta$, hence $G > -1 + \delta$.

(ii) If two are positive and two are negative, assume $z_1, z_4 > 0, z_2, z_3 < 0$. If $\frac{z_j}{z_i} = -2$ for some i, j, let $z_1 = u > 0, z_2 = -2u; z_3 = -v < 0, z_4 = u + v$ where u, v are coprime. Then

$$G = -\frac{1}{2} - \frac{1}{uv} + \frac{1}{u(u+v)} + \frac{(2, v)^2}{2uv} - \frac{(2, u+v)^2}{2u(u+v)} - \frac{1}{v(u+v)},$$

$$G \geqslant -\frac{1}{2} - \frac{1}{uv} + \frac{1}{u(u+v)} + \frac{1}{2uv} - \frac{4}{2u(u+v)} - \frac{1}{v(u+v)}$$

$$= -\frac{1}{2} - \frac{3}{2uv} > -1 + \delta.$$

The last step requires $(u, v) \neq (1, 1), (1, 2), (2, 1), (1, 3), (3, 1)$. We can easily check that $(1, 1)$ gives $z_1 + z_3 = 0, (1, 2)$ gives $G = 0$, and the others give golden ratios. So, $G > -1 + \delta$.

If $\frac{z_j}{z_i} = -3$ for some i, j, let $z_1 = u > 0, z_2 = -3u; z_3 = -v < 0, z_4 = 2u + v$ where u, v are coprime. Then

$$G = -\frac{1}{3} - \frac{1}{uv} + \frac{1}{u(2u + v)} + \frac{(3, v)^2}{3uv} - \frac{(3, 2u + v)^2}{3u(2u + v)} - \frac{(2, v)^2}{v(2u + v)},$$

$$G \geqslant -\frac{1}{3} - \frac{1}{uv} + \frac{1}{u(2u + v)} + \frac{1}{3uv} - \frac{9}{3u(2u + v)} - \frac{4}{v(2u + v)}$$

$$= -\frac{1}{3} - \frac{8}{3uv} > -1 + \delta.$$

The last step requires $uv > 4$. Otherwise, $(u, v) = (1, 1), (1, 2), (1, 4)$ give opposite numbers or golden ratios, and $(u, v) = (2, 1), (1, 3), (3, 1), (4, 1)$ give $G = -\frac{4}{5}, \frac{2}{5}, -\frac{2}{3}, -\frac{2}{3}$, respectively. So $G > -1 + \delta$.

If $\frac{z_j}{z_i} \neq -2, -3$ for every i, j, then the four negative terms in G are all $\geqslant -\frac{1}{4}$ and the two positive terms are larger than δ, and we still have $G > -1 + \delta$.

Now we have verified $L_0 + L_1 \geqslant \lfloor \frac{p}{12} \rfloor$ and thereby the problem statement.

Remark 1 The idea for question (1) is from the trace of the points $\left(\left\{ \frac{ak}{m} \right\}, \left\{ \frac{bk}{m} \right\} \right)$ in the unit square.

For each point (x, y), consider the value of xy. In a rectangle whose sides are parallel to the coordinate axes, the average xy value of all points is equal to the xy value at the center; the average xy value along a diagonal is equal to the xy value at the center plus $\frac{1}{12}$ times the area of the rectangle.

In this way, our original view of the xy values of a series of discrete points is transferred to several line segments, and then by splicing and regrouping them into ab segments, we focus on the corresponding ab small rectangles, which together form a unit square.

Remark 2 In question (1), the estimate of ③ uses Gauss' trapezium rule, which causes error of order $\frac{1}{m^2}$. Another approach is to compare $\left\{ \frac{ak}{m} \right\} \left\{ \frac{bk}{m} \right\}$ with $\int_{k-\frac{1}{2}}^{k+\frac{1}{2}} \left\{ \frac{ax}{m} \right\} \left\{ \frac{bx}{m} \right\} dx$, which leads to similar results. However, if one simply compares it with the integral from k to $k + 1$, then the error will involve terms with ab, and this will make further discussion impossible.

Test I, Second Day
(8 am–12:30 pm; 12 March 2023)

4 For positive integers m, n, define

$$S(m, n) = \{(a, b) \in \mathbf{Z}_+^2 \mid 1 \leqslant a \leqslant m, 1 \leqslant b \leqslant n, \gcd(a, b) = 1\}.$$

Prove that for any positive integers d and r, there exist integers $m, n \geqslant d$, such that

$$|S(m, n)| \equiv r \pmod{d}.$$

Here, $|A|$ represents the number of elements of a finite set A.

(Contributed by Fu Yunhao)

Solution   Let $n = d + r, m = d \cdot (d + r)! + 1$. For $1 \leqslant b \leqslant d + r$, the number of integers $1, 2, \ldots, m$ that are coprime with b is

$$\frac{\varphi(b)}{b} \cdot d \cdot (d + r)! + 1.$$

Note that $b \mid (d+r)!$, and thus the above number is equivalent to 1 modulo d. We have,

$$|S(m, n)| \equiv \sum_{b=1}^{d+r} 1 \equiv r \pmod{d}.$$

5 In triangle ABC, $P_1, \ldots, P_n$ are n interior points such that no three points of $P_1, \ldots, P_n, A, B$ and C are collinear. Prove that triangle ABC can be split into $2n + 1$ small triangles such that every vertex of them is from $P_1, \ldots, P_n, A, B, C$, and the number of small triangles with one or more vertex at A, B or C is at least $n + \sqrt{n} + 1$.

(Contributed by Ai Yinghua)

Solution   For side BC, we define a partial order $<_A$ in the set $P = \{P_1, \ldots, P_n\}$:

$P_i <_A P_j \Leftrightarrow P_i$ is an interior point of triangle $P_j BC \Leftrightarrow$ the ray $P_j P_i$ intersects side BC.

Similarly for side CA, define partial order $<_B$ in P. If distinct points P_i, P_j are incomparable under $<_A$, then line $P_i P_j$ does not intersect segment BC. As $P_i P_j$ intersects triangle ABC at two points, we infer that P_i, P_j are comparable under $<_B$, and therefore an antichain under the partial order $<_A$ is a chain under the partial order $<_B$.

According to Dilworth theorem, under the partial order $<_A$, the n-set P contains either a chain or an antichain of length at least $\sqrt{n}$. If P does not contain one such chain, then as we have shown, the antichain under $<_A$ is a chain under $<_B$. Hence, there is always a chain of length at least $\sqrt{n}$: either under $<_A$, or under $<_B$.

By symmetry, suppose that under $<_A$ there is a chain of length at least $\sqrt{n}: P_1 <_A \cdots <_A P_t$, where $t \geqslant \sqrt{n}$. Draw segments

$$BP_i, CP_i(1 \leqslant i \leqslant t), P_1P_2, \ldots, P_{t-1}P_t, P_tA,$$

which split triangle ABC into $2t + 1$ triangles

$$\triangle BCP_1, \quad \triangle BP_iP_{i+1}, \quad \triangle CP_iP_{i+1} \quad (1 \leqslant i \leqslant t-1), \quad \triangle BP_tA, \quad \triangle CP_tA. \tag{$*$}$$

Let $k_0, k_1, \ldots, k_t$ be the numbers of P-points in $\triangle BCP_1, \triangle BP_1P_2, \ldots,$ $\triangle BP_tA$, respectively. For each such triangle, say $\triangle BQR$, let $U_1, \ldots, U_k$ be the P-points in it, assuming $\angle QBU_j$ $(1 \leqslant j \leqslant k)$ in the increasing order. We draw segments BU_j $(1 \leqslant j \leqslant k)$ and $QU_1, U_1U_2, \ldots, U_kR$ to obtain $k + 1$ small triangles all with vertex B, and then split the polygon $QU_1 \cdots U_kR$ into k small triangles (it is well known such triangulation always exists). Now $\triangle BQR$ is split into $2k+1$ small triangles, and $k+1$ of them have B as a vertex. Likewise, if $l_1, \ldots, l_t$ are the numbers of P-points in $\triangle CP_1P_2, \ldots, \triangle CP_tA$, respectively, then each triangle can be split into $2l + 1$ small triangles, and $l + 1$ of them have C as a vertex.

So we have constructed a triangulation of ABC, in which at least

$$T = \sum_{i=0}^{t}(k_i + 1) + \sum_{i=1}^{t}(l_i + 1)$$

small triangles have B or C as a vertex. Since $k_0 + k_1 + \cdots + k_t + l_1 + \cdots + l_t = n - t$, we reach

$$T = (n - t) + (2t + 1) = n + t + 1 \geqslant n + \sqrt{n} + 1,$$

which is the desired conclusion.

6 (1) Prove that on the complex plane, on each line through the origin except the real axis, there is at most one point z satisfying that $\frac{1+z^{23}}{z^{64}}$ is a real number.

(2) Prove that for any complex number $a \neq 0$ and any real number θ, the equation $1 + z^{23} + az^{64} = 0$ has at least one root in the set

$$S_\theta = \left\{ z \in \mathbf{C} \,\middle|\, \mathrm{Re}(z \cdot e^{-i\theta}) \geqslant |z| \cdot \cos \frac{\pi}{20} \right\}.$$

(Contributed by Yao Yijun)

Solution   We assume the angles are modulo 2π in the context.

(1) For non-zero complex number z, let $z = r(\cos\theta + i\sin\theta)(r > 0, \theta \in [0, 2\pi))$. Define

$$f(z) := \frac{1 + z^{23}}{z^{64}}$$

$$= \frac{1}{r^{64}} [(\cos(-64\theta) + i\sin(-64\theta)) + r^{23}(\cos(-41\theta) + i\sin(-41\theta))],$$

and we have

$$\frac{1 + z^{23}}{z^{64}} \in \mathbf{R} \Leftrightarrow \mathrm{Im}\frac{1 + z^{23}}{z^{64}} = \frac{-1}{r^{64}}(\sin 64\theta + r^{23}\sin 41\theta) = 0$$

$$\Leftrightarrow \sin 64\theta + r^{23}\sin 41\theta = 0.$$

For $\theta \neq 0, \pi$, if $\sin 41\theta = 0$, then $\sin 64\theta \neq 0$, the above equation does not have a solution.

If $\sin 41\theta \neq 0$, then

$$r = g_0(\theta) := \left(\frac{-\sin 64\theta}{\sin 41\theta} \right)^{\frac{1}{23}}$$

is uniquely determined. To get $r > 0$, $\sin 64\theta$ and $\sin 41\theta$ must have opposite signs. Note that $g_0(\theta) = -g_0(\theta + \pi)$, and the conclusion follows.

(2) To start, note that

$$S_\theta = \left\{ re^{it} : r \geqslant 0, \theta - \frac{\pi}{20} \leqslant t \leqslant \theta + \frac{\pi}{20} \right\}$$

is a sector centered at the origin. We think about the problem statement in reverse: if as z traverses all nonzero elements of S_θ, $a = -\frac{1 + z^{23}}{z^{64}}$ takes every complex number, then the equation $1 + z^{23} + az^{64} = 0$ always has a solution in S_θ.

Use the notations as in Solution 1, and let $a = \rho e^{i\lambda}$ ($\rho > 0, \lambda \in [0, 2\pi)$). We have

$$f(z) = -a = -\rho e^{i\lambda}.$$

For fixed $\lambda \in [0, \pi)$, consider the set of complex numbers

$$\Gamma_\lambda : \{z | f(z)e^{-i\lambda} \in \mathbf{R}\} = \{z | \sin(\lambda + 64\theta) + r^{23}\sin(\lambda + 41\theta) = 0\}.$$

There are two situations:

(1°) $\lambda + 64\theta \equiv \lambda + 41\theta \equiv 0 \pmod{\pi}$, implying that

$$\theta \equiv \frac{k\pi}{23}, \quad \lambda \equiv \frac{5k\pi}{23} \pmod{\pi}$$

for some integer k.

(2°) The congruence equations in (1°) do not hold. By an argument similar to Solution 1, we see that on each line through the origin, there is at most one point z that belongs to Γ_λ; in particular, any ray starting at the origin intersects Γ_λ at one point at most.

The set Γ_λ consists of curves described by polar equations. Define

$$g_\lambda(\theta) := \left(\frac{-\sin(\lambda + 64\theta)}{\sin(\lambda + 41\theta)} \right)^{\frac{1}{23}}.$$

As said, we expect that when z lies in a sector region where the range of the angles depends on θ (a θ-interval), and in this set such that $g_\lambda(\theta) > 0$,

$$-e^{-i\lambda}f(z) = -\frac{1}{(g_\lambda(\theta))^{64}}(\cos(\lambda + 64\theta) + r^{23}\cos(\lambda + 41\theta))$$

traverses all real numbers.

$$(**)$$

The θ values with $g_\lambda(\theta) > 0$ consist of several subintervals. At the endpoints of these intervals, there are two possible cases:

(a) The endpoint θ_1 satisfies $\lambda + 41\theta_1 \equiv 0 \pmod{\pi}$, yet $\lambda + 64\theta_1 \not\equiv 0 \pmod{\pi}$. As $\theta \to \theta_1$ in the interval, $g_\lambda(\theta) \to \infty$ and correspondingly $a \to 0(\rho \to 0)$;

(b) The endpoint θ_1 satisfies $\lambda + 64\theta_1 \equiv 0 \pmod{\pi}$, yet $\lambda + 41\theta_1 \not\equiv 0 \pmod{\pi}$. As $\theta \to \theta_1$ in the interval, $g_\lambda(\theta) \to 0$ and correspondingly $a \to \infty(\rho \to +\infty$ or $-\infty$ if $\cos(\lambda + 64\theta_1) = -1$ or 1, respectively).

Now we claim that for any θ-interval of length $\frac{\pi}{10}$, there is a subinterval in which $(**)$ holds.

(1) If for certain λ, there exists θ_0 in the interval satisfying condition (1°), then the distance from one of the endpoints, say the right one to θ_0 is $\geqslant \frac{\pi}{20} > \frac{2\pi}{41} > \frac{3\pi}{64}$. Consequently, in the process of θ_0 approaching this endpoint, two possible situations may occur (without loss of generality,

assume $\sin(\lambda + 64\theta) > 0$ in $\left(\theta_0, \theta_0 + \frac{\pi}{64}\right)$): one is shown in the following table:

	θ_0		$\theta_0+\frac{\pi}{64}$		$\theta_0+\frac{\pi}{41}$		$\theta_0+\frac{\pi}{32}$		$\theta_0+\frac{3\pi}{64}$		$\theta_0+\frac{2\pi}{41}$
$\sin(\lambda+64\theta)$	0	+	0	−	−	−	0	+	0	−	−
$\sin(\lambda+41\theta)$	0	−	−	−	0	+	+	+	+	+	0
$\cos(\lambda+64\theta)$	±1		∓1				±1		∓1		

We see the piecewise continuous function $\rho(\theta)$ satisfies

$$\lim_{\theta \to \left(\theta_0+\frac{\pi}{32}\right)^-} \rho = \pm\infty, \qquad \lim_{\theta \to \left(\theta_0+\frac{3\pi}{64}\right)^+} \rho = \mp\infty,$$

$$\rho\left(\theta_0 + \frac{\pi}{41}\right) = \rho\left(\theta_0 + \frac{2\pi}{41}\right) = 0,$$

and hence on $\left[\theta_0 + \frac{\pi}{41}, \theta_0 + \frac{\pi}{32}\right) \bigcup \left(\theta_0 + \frac{3\pi}{64}, \theta_0 + \frac{2\pi}{41}\right]$, ρ can attain every real value. The other possible situation is shown in the following table:

	θ_0		$\theta_0+\frac{\pi}{64}$		$\theta_0+\frac{\pi}{41}$		$\theta_0+\frac{\pi}{32}$		$\theta_0+\frac{3\pi}{64}$		$\theta_0+\frac{2\pi}{41}$
$\sin(\lambda+64\theta)$	0	+	0	−	−	−	0	+	0	−	−
$\sin(\lambda+41\theta)$	0	+	+	+	0	−	−	−	−	−	0
$\cos(\lambda+64\theta)$	±1		∓1				±1		∓1		

We see that ρ attains every real value on $\left(\theta_0 + \frac{\pi}{32}, \theta_0 + \frac{3\pi}{64}\right)$.

(2) If θ, λ satisfy condition (2°). We observe that this θ-interval contains at least four zeros of $\sin(\lambda + 41\theta)$. As $\frac{3}{43} > \frac{4}{64}$, there must be two adjacent zeros of $\sin(\lambda + 41\theta)$, say θ_1 and $\theta_1 + \frac{\pi}{41}$, in between there are two zeros of $\sin(\lambda + 64\theta)$, say θ_2 and $\theta_2 + \frac{\pi}{64}$. Assuming that $\theta_1 + \frac{2\pi}{41}$ also lies in this interval, we have either the following table (without loss of generality, let $\sin(\lambda + 64\theta) > 0$ in (θ_1, θ_2)),

	θ_1		θ_2		$\theta_2+\frac{\pi}{64}$		$\theta_1+\frac{\pi}{41}$		$\theta_2+\frac{\pi}{32}$
$\sin(\lambda+64\theta)$	+	+	0	−	0	+	+	+	0
$\sin(\lambda+41\theta)$	0	−	−	−	−	−	0	+	+
$\cos(\lambda+64\theta)$			±1		∓1				±1

and taking $[\theta_1, \theta_2) \bigcup \left(\theta_2 + \frac{\pi}{64}, \theta_1 + \frac{\pi}{41}\right]$ suffices; or the following table:

	θ_1		θ_2		$\theta_2 + \frac{\pi}{64}$		$\theta_1 + \frac{\pi}{41}$		$\theta_2 + \frac{\pi}{32}$
$\sin(\lambda + 64\theta)$	+	+	0	−	0	+	+	+	0
$\sin(\lambda + 41\theta)$	0	+	+	+	+	+	0	−	−
$\cos(\lambda + 64\theta)$			±1		∓1				±1

and taking $\left(\theta_2, \ \theta_2 + \frac{\pi}{64}\right]$ suffices.

In summary, the problem statement is verified.

Remark 1 This type of result traces back to P. Nekrasoff (*Ueber trinomische Gleichugen, Math. Annalen*, 29(1887), pp. 413–430). The proof idea comes from Polish mathematician M. Biernacki's doctoral thesis Sur les équations algébriques contenant des paramètres arbitraires (*Bulletin de l'Académie polonaise des Sciences et des Lettre, classe des Sciences Mathématiques, Série A*, 1927, pp. 541–685).

Remark 2 Using complex analysis theory, one can reduce $\frac{\pi}{20}$ to $\frac{\pi}{32}$.

Test II, First Day
(8 am–12:30 pm; 16 March 2023)

1 Given positive integers n and a satisfying $n > 1$, $(n, a) = 1$. A country consists of n islands $D_1, D_2, \ldots, D_n$. For any two different islands D_i and D_j, there is a one-way ferry from D_i to D_j if and only if

$$ij \equiv ia \pmod{n}.$$

A tourist wants to visit as many islands as possible. He can fly to any island that he chooses and start the tour, but thereafter he must use the one-way ferries to travel between the islands. Find the maximum number of islands that the tourist can possibly visit.

(Contributed by Qu Zhenhua)

Solution For prime number p, define $v_p(n)$ as the exponent of p in the factorization of n. Let $x(n) = \sum_{v_p(n) \geqslant 2} 1$, $y(n) = \sum_{v_p(n)=1} 1$. We will prove that the maximum number of islands that the tourist can possibly visit, $k(n)$, is given by

$$k(n) = \begin{cases} 3x(n) + 2y(n) + 1, & v_2(n) \neq 1, \\ 3x(n) + 2y(n), & v_2(n) = 1. \end{cases}$$

For $x, y \in \mathbf{Z}, m \in \mathbf{Z}_{>0}$, if $xy \equiv xa \pmod{m}$, then write $x \to y \pmod{m}$. Let p be a prime factor of n, $p^\alpha \parallel n$. For sequence $x_1 \to x_2 \to \cdots \to x_k \pmod{n}$, it remains the same when modulo p^α:

$$x_1 \to x_2 \to \cdots \to x_k \pmod{p^\alpha}.$$

(1) If $x_i \not\equiv 0 \pmod{p^\alpha}$, then x_{i+1} and p are coprime. This is because $p^\alpha \mid x_i(x_{i+1} - a)$, $p^\alpha \nmid x_i$, and thus $p \mid (x_{i+1} - a)$; as a, n are coprime, we have $p \nmid a$, and so $p \nmid x_{i+1}$.

(2) If x_i and p are coprime, then $x_{i+1} \equiv a \pmod{p^\alpha}$. This is because $p^\alpha \mid x_i(x_{i+1} - a)$, as $p \nmid x_i$, we have $p^\alpha \mid (x_{i+1} - a)$.

(3) From (1) and (2), we see that if $\alpha \geqslant 2$, the values $x_1, x_2, \ldots, x_k \pmod{p^\alpha}$ change at most three times: from 0 to a nonzero number modulo p^α, then to a number coprime with p, then to a modulo p^α.

If $\alpha = 1, p \neq 2$, then $x_1, x_2, \ldots, x_k \pmod{p^\alpha}$ change at most twice: from 0 modulo p to a number coprime with p, then to a modulo p.

If $\alpha = 1, p = 2$, then $x_1, x_2, \ldots, x_k \pmod{p^\alpha}$ change at most once: from 0 modulo 2 to a modulo 2.

Suppose that any adjacent pair in the sequence $x_1, x_2, \ldots, x_k \pmod{n}$ are different. Then in the sequence, each following number differs from the preceding number modulo p^α for some p, and therefore $k \leqslant k(n)$.

(4) Now we construct examples with $k = k(n)$. First, for modulo p^α:

  (i) If $\alpha \geqslant 2$, then $0 \to p \to a + p^{\alpha-1} \to a \pmod{p^a}$;

 (ii) If $\alpha = 1, p \neq 2$, then $0 \to b \to a \pmod{p}$, where b, p are coprime and $b \not\equiv a \pmod{p}$;

(iii) If $\alpha = 1, p = 2$, then $0 \to a \pmod{2}$.

Let $n = p_1^{\alpha_1} \cdots p_w^{\alpha_w}$ be the prime factorization of n. For each p^α, take a sequence that obeys (i)–(iii). As $0 \to 0, a \to a$ are allowed, we can add a few 0's at the beginning and a few a's at the end in order to get a sequence of length $k(n)$, denoted by L_i. We must require that for $p_i^{\alpha_i}$ and $p_j^{\alpha_j} (i \neq j)$, the places where the respective numbers change in L_i and L_j are different.

For example, $\alpha_1 \geqslant 2$, we take

$$L_1 : 0 \to p_1 \to a + p_1^{\alpha_1 - 1} \to a \to a \to \cdots \to a \pmod{p_1^{\alpha_1}};$$

then $\alpha_2 = 1, p_2 \neq 2$, and take

$$L_2 : 0 \to 0 \to 0 \to 0 \to b \to a \to a \to \cdots \to a \pmod{p_2};$$

and so on for $L_3, \ldots, L_w$.

The value of x_t is determined by the tth terms of $L_1, L_2, \ldots, L_w$ and Chinese remainder theorem. As $x_t \to x_{t+1} \pmod{p_i^{\alpha_i}}, i = 1, 2, \ldots, w$, we have $x_t \to x_{t+1} \pmod{n}$. Finally, since $x_1, x_2, \ldots, x_{k(n)}$ modulo n are distinct, the maximum number $k(n)$ (given at the beginning of the solution) can be attained. This completes the proof.

2 Let ABC be a scalene acute-angled triangle, where AP, BQ, CR are the altitudes and H is the orthocenter. Let the line passing through A and parallel to BC intersect line RQ at point D. Let A_1 be the midpoint of BC; K be the point of intersection of RQ and AA_1. Suppose that the line through the midpoint of AH and point K meets line DA_1 at point A_2; similarly, define points B_2 and C_2.

Let ω be the circumcircle of non-degenerate triangle $A_2 B_2 C_2$. Prove that there exist three circles $\odot A', \odot B'$ and $\odot C'$ inscribed in ω that satisfy the following conditions:

(1) $\odot A'$ is tangent to sides AB, AC; $\odot B'$ is tangent to sides BA, BC; $\odot C'$ is tangent to sides CA, CB;
(2) The centers A', B', C' are distinct and collinear.

(Contributed by Zhang Sihui)

Solution 1 Let the midpoint of AH be A^*. As illustrated in Fig. 7.3, $\angle A_1 RH = \angle A_1 CR = \angle RAH$, implying that $A_1 R$ is a tangent line of the circle say Γ with diameter AH. Likewise, $A_1 Q$ is also a tangent line of circle Γ (R, Q as the points of tangency), and thus

$$A_1 A^* \perp RQ. \qquad \textcircled{6}$$

Consider the polar reciprocation in circle Γ: D lies on the polar line of A_1, and conversely A_1 lies on the polar line of D in Γ. But DA is a tangent

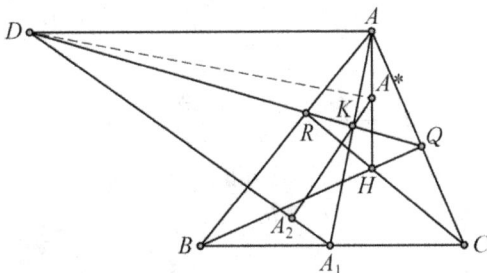

Fig. 7.3

of Γ, and thus AA_1 is the polar line of D, giving

$$DA^* \perp AA_1. \qquad \qquad ⑦$$

From ⑥ and ⑦, it follows that K is the orthocenter of $\triangle A^* A_1 D$, and in particular, $A^* K \perp A_1 D$, yielding $\angle A^* A_2 A_1 = 90°$. Therefore, A_2 lies on the nine-point circle of $\triangle ABC$, which is just circle ω.

As shown in Fig. 7.4, let N be the center of ω and I be the incenter of $\triangle ABC$. Let $BC = a, CA = b, AB = c$, and $s = \frac{a+b+c}{2}$. We introduce the inversion transformation f with A as the center and the nine-point circle invariant. Suppose that $\odot A'$ is the image of the incircle under f. Clearly, $\odot A'$ is tangent to sides AB and AC. Define $\odot B'$ and $\odot C'$ in a similar way.

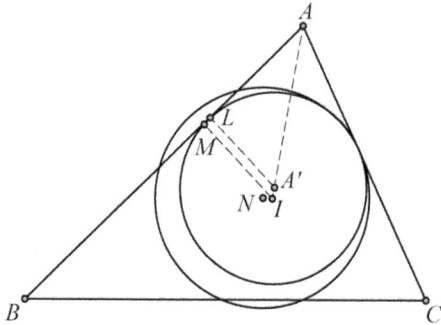

Fig. 7.4

Let M be the foot of the perpendicular dropped from I to side AB. We assert that $\odot A'$ does not coincide with ω, as otherwise $AR \cdot AC_1 = AM^2$, or $b \cdot \cos A \cdot \frac{c}{2} = \left(\frac{b+c-a}{2}\right)^2$, implying that $a = b$ or $a = c$, which contradicts the assumption that $\triangle ABC$ is not isosceles.

In the following, we verify A', B', C' are collinear. Suppose that

$$\overrightarrow{IA'} = p \cdot \overrightarrow{IA}, \overrightarrow{IB'} = q \cdot \overrightarrow{IB}, \overrightarrow{IC'} = r \cdot \overrightarrow{IC}.$$

Then

$$\frac{a}{p} \cdot \overrightarrow{IA'} + \frac{b}{q} \cdot \overrightarrow{IB'} + \frac{c}{r} \cdot \overrightarrow{IC'} = a \cdot \overrightarrow{IA} + b \cdot \overrightarrow{IB} + c \cdot \overrightarrow{IC} = \overrightarrow{0}.$$

To verify A', B', C' are collinear, it suffices to show

$$\frac{a}{p} + \frac{b}{q} + \frac{c}{r} = 0.$$

Let L be the foot of the altitude from A' to side AB. Then $\frac{AA'}{AI} = \frac{AL}{AM}, AM = s - a$. By properties of the inversion f, we see $f(M) = L$, and hence

$$AM \cdot AL = AR \cdot AC_1 = b \cos A \cdot \frac{c}{2} = \frac{b^2 + c^2 - a^2}{4},$$

$$\frac{AL}{AM} = \frac{AM \cdot AL}{AM^2} = \frac{b^2 + c^2 - a^2}{4(s-a)^2},$$

yielding that

$$p = \frac{IA'}{IA} = \frac{2(a-b)(a-c)}{(b+c-a)^2},$$

and analogously

$$q = \frac{2(b-c)(b-a)}{(c+a-b)^2}, r = \frac{2(c-a)(c-b)}{(a+b-c)^2}.$$

Finally, the above equations give $\frac{a}{p} + \frac{b}{q} + \frac{c}{r} = 0$. (It requires to find the cyclic sum of $a(b-c)(b+c-a)^2$. Treat it as a function of a, $g(a)$. Note that the coefficient of a^3 vanishes, so $g(a)$ is quadratic, with $g(b) = g(c) = g(b+c) = 0$, and hence $g \equiv 0$.)

Solution 2 Similar to Solution 1, we see that ω is the nine-point circle of $\triangle ABC$. To handle a geometry problem with angle bisectors in a circumcircle, a common method is to treat the circumcircle as the unit circle in the complex plane and let

$$A = a^2, \quad B = b^2, \quad C = c^2.$$

The midpoints of the minor arcs $\overparen{BC}, \overparen{CA}, \overparen{AB}$ are $-bc, -ca, -ab$, respectively; the incenter $I = -(ab+bc+ca)$; the nine-point center $N = \frac{a^2+b^2+c^2}{2}$.

Let A' correspond to complex number a'. As A' lies on the bisector of $\angle CAB$ (the ray starting from A and passing through I), there exists $\lambda > 0$ such that

$$a' = a^2 + \lambda(-(ab+bc+ca) - a^2) = a^2 - \lambda(a+b)(a+c).$$

Moreover, $\odot A'$ is inscribed in ω (since A' is inside ω) and tangent to sides AB, AC. It follows that

$$|NA'| + |AA'| \cdot \sin\frac{A}{2} = \frac{1}{2}$$

$$\Leftrightarrow \left| \frac{a^2+b^2+c^2}{2} - a^2 - \lambda(a+b)(a+c) \right| = \frac{1}{2} - \lambda|a+b| \cdot |a+c| \cdot \frac{|b^2+bc|}{2}$$

$$\Leftrightarrow |(-a^2+b^2+c^2) - 2\lambda(a+b)(a+c)|^2 = (1 - \lambda|a+b| \cdot |a+c| \cdot |b+c|)^2.$$

This is a quadratic equation in λ. After simplification, it turns out that the quadratic coefficient is

$$4|a+b|^2 \cdot |a+c|^2 - |a+b|^2 \cdot |a+c|^2 \cdot |b+c|^2$$
$$= |a+b|^2 \cdot |a+c|^2 \cdot |b-c|^2, \qquad (*)$$

while the constant term is $|-a^2+b^2+c^2|^2 - 1$. According to Feuerbach theorem, one root is $\lambda = 1$, and the other root is the constant term divided by $(*)$. Consequently,

$$a' = a^2 - \lambda(a+b)(a+c)$$

$$= a^2 - \frac{(a+b)(a+c)}{|a+b|^2 \cdot |a+c|^2 \cdot |b-c|^2}[|-a^2+b^2+c^2|^2 - 1]$$

$$= a^2 - \frac{1}{(\bar{a}+\bar{b})\cdot(\bar{a}+\bar{c})\cdot(b-c)(\bar{b}-\bar{c})}$$
$$\times [(-a^2+b^2+c^2)(-\bar{a}^2+\bar{b}^2+\bar{c}^2) - 1]$$

$$= a^2 - \frac{ab \cdot ac \cdot (-bc)}{(a+b)\cdot(a+c)\cdot(b-c)^2}$$
$$\times \left[(-a^2+b^2+c^2)\left(-\frac{1}{a^2}+\frac{1}{b^2}+\frac{1}{c^2}\right) - 1\right]$$

$$= a^2 - \frac{-a^2b^2c^2}{(a+b)\cdot(a+c)\cdot(b-c)^2}$$
$$\cdot \frac{(-a^2+b^2+c^2)(-b^2c^2+a^2c^2+a^2b^2) - a^2b^2c^2}{a^2b^2c^2}$$

$$= a^2 + \frac{a^2b^2c^2 - b^2c^2(b^2+c^2) - a^4(c^2+b^2) + a^2(b^2+c^2)^2 - a^2b^2c^2}{(a+b)\cdot(a+c)\cdot(b-c)^2}$$

$$= a^2 + \frac{(-a^4+a^2(b^2+c^2)-b^2c^2)(b^2+c^2)}{(a+b)\cdot(a+c)\cdot(b-c)^2}$$

$$= a^2 - \frac{(a^2-b^2)(a^2-c^2)(b^2+c^2)}{(a+b)\cdot(a+c)\cdot(b-c)^2}$$

$$= a^2 - \frac{(a-b)(a-c)(b^2+c^2)}{(b-c)^2}. \qquad (**)$$

As we know,

$$a', b', c' \text{ are collinear} \Leftrightarrow a'-I, b'-I, c'-I \text{ are collinear}.$$

By (∗∗), it follows that

$$a' - I = (a+b)(a+c) - \frac{(a-b)(a-c)(b^2+c^2)}{(b-c)^2}$$

$$= \frac{(a+b)(a+c)(b-c)^2 - (a-b)(a-c)(b^2+c^2)}{(b-c)^2}$$

$$= \frac{(a+b)(a+c)(-2bc) + [(a+b)(a+c) - (a-b)(a-c)](b^2+c^2)}{(b-c)^2}$$

$$= \frac{(a+b)(a+c)(-2bc) + 2a(b+c)(b^2+c^2)}{(b-c)^2}$$

$$= \frac{-2(bca^2 - (b^3+c^3)a + b^2c^2)}{(b-c)^2}$$

$$= \frac{-2(ab-c^2)(ca-b^2)}{(b-c)^2}.$$

Now, we need only show the three points

$$X = \frac{(ab-c^2)(ca-b^2)}{(b-c)^2}, \quad Y = \frac{(bc-a^2)(ab-c^2)}{(c-a)^2}, \quad Z = \frac{(ca-b^2)(bc-a^2)}{(a-b)^2}$$

are collinear. By plane geometry in complex methods, it is equivalent to

$$0 = \begin{vmatrix} X & Y & Z \\ \overline{X} & \overline{Y} & \overline{Z} \\ 1 & 1 & 1 \end{vmatrix} = \begin{vmatrix} X & Y & Z \\ \frac{X}{a^2bc} & \frac{Y}{ab^2c} & \frac{Z}{abc^2} \\ 1 & 1 & 1 \end{vmatrix} = \frac{1}{abc} \begin{vmatrix} X & Y & Z \\ \frac{X}{a} & \frac{Y}{b} & \frac{Z}{c} \\ 1 & 1 & 1 \end{vmatrix}$$

$$= \frac{1}{abc}\left(\left(\frac{1}{b} - \frac{1}{a}\right) XY - \left(\frac{1}{c} - \frac{1}{a}\right) XZ + \left(\frac{1}{c} - \frac{1}{b}\right) YZ \right)$$

$$= \frac{1}{abc}\left(\frac{a-b}{ab} XY + \frac{b-c}{bc} YZ - \frac{a-c}{ca} ZX \right)$$

$$= \frac{\prod_{cyc}(ab-c^2)}{a^2b^2c^2}$$

$$\times \left(\frac{c(a-b)(ab-c^2)}{(b-c)^2(c-a)^2} + \frac{a(b-c)(bc-a^2)}{(c-a)^2(a-b)^2} + \frac{b(c-a)(ca-b^2)}{(b-c)^2(a-b)^2} \right)$$

$$= \frac{\prod_{cyc}(ab-c^2)}{a^2b^2c^2(a-b)^2(b-c)^2(c-a)^2} \sum_{cyc} a(b-c)^3(bc-a^2).$$

Finally, we focus on the cyclic sum

$$\sum_{cyc} a(b-c)^3(bc-a^2) = abc\sum_{cyc}(b-c)^3 - \sum_{cyc}a^3(b-c)^3$$

$$= abc\left\{\sum_{cyc}(b^3-c^3) - 3\sum_{cyc}(b^2c-bc^2)\right\}$$

$$-\left\{\sum_{cyc}(a^3b^3-a^3c^3) - 3abc\sum_{cyc}(a^2b-a^2c)\right\} = 0,$$

as all four cyclic sums in the third row cancel out. This shows the determinant equals zero and the three centers are collinear.

3 Find the largest positive integer m, such that one can possibly color some squares of the 70×70 grid red satisfying that

(1) there are no two red squares that have the same number of red squares in their rows and the same number of red squares in their columns;

(2) there are at least two rows, each with exactly m red squares.

(Contributed by Fu Yunhao)

Solution To start, suppose there are a red squares in the row with maximum number of red squares (say the first row), and b red squares in the column with maximum number of red squares (say the first column). If $a > b$, consider the columns of those red squares in the first row: they must contain distinct numbers of red squares, and these numbers come from $1, 2, \ldots, b$, which have fewer than a choices, a contradiction. Similarly, $a < b$ is impossible, and thus $a = b$. Moreover, the numbers of red squares in the columns where the red squares in the first row are located are exactly a permutation of $1, 2, \ldots, a$, and the same for the numbers of red squares in the rows where the red squares in the first column are located as well. This means for every $1 \leqslant i \leqslant a$, there is always a row and a column each with exactly i red squares.

We claim: there exists a coloring of the 70×70 grid satisfying (1) and for which exactly x_i rows each contain i red squares, exactly y_i columns each contain i red squares $(x_i, y_i \in \mathbf{Z}_{>0}, i = 1, 2, \ldots, a)$, if and only if there is a coloring of an $a \times a$ table, such that its ith row contains exactly ix_i red squares and its jth column contains exactly iy_i red squares $(i = 1, 2, \ldots, a)$. (*)

First, we prove the necessity. If a coloring of the 70×70 grid P satisfies (1), then construct an $a \times a$ table Q such that for every $i, j \in \{1, 2, \ldots, a\}$, the ith row, jth column of Q is colored red if and only if there is a red square in P whose row and column contain i and j red squares, respectively (call this condition $T(i, j)$). Note that in P, the number of red squares whose rows contain i red squares is ix_i, and the columns where they are located have distinct numbers of red squares. Therefore, the number of j's satisfying condition $T(i, j)$ is exactly ix_i, indicating the ith row of Q contains ix_i red squares. For the same reason, the ith column of Q contains iy_i red squares, and the necessity is verified.

Next, we prove the sufficiency. Suppose in an $a \times a$ table Q, the ith row contains exactly ix_i red squares and the jth column contains exactly iy_i red squares. Let us color the 70×70 grid P as follows. Label the rows $R(i, x)(1 \leqslant x \leqslant x_i)$ and the columns $C(i, y)(1 \leqslant y \leqslant y_i)$. If the ith row, jth column of Q is red, call (i, j) a "good" pair. For a good pair (i_0, j_0), consider all good pairs (i_0, j) and let j_0 be the $v(i_0, j_0)$th smallest of them; also, consider all good pairs (i, j_0) and let i_0 be the $u(i_0, j_0)$th smallest of them. Now we color the square in row $R\left(i_0, \left\lfloor \frac{v(i_0, j_0)}{i_0} \right\rfloor\right)$ and column $C\left(j_0, \left\lfloor \frac{u(i_0, j_0)}{j_0} \right\rfloor\right)$ of P red. In this way, every good pair (i_0, j_0) corresponds to only one pair of row $R(i_0, x)$ and column $C(j_0, y)$ (the intersecting square is colored), and hence condition (1) is met; conversely, for each row $R(i_0, x)$, the red squares in it have exactly the j's such that (i_0, j_0) is a good pair and they are ranked $(x-1)i_0+1, (x-1)i_0+2, \ldots, xi_0$ in the increasing sequence of them. For the same reason, each column $C(j_0, y)$ contains exactly j_0 red squares. The sufficiency is verified.

We return to the original problem and give a coloring scheme with $m = 32$. In the 64×64 table Q, color the square in row i, column j if and only if $i + j \leqslant 66$. In this way, the numbers of red squares in the rows and in the columns of Q are both $64, 64, 63, 62, \ldots, 3, 2$. Now let

$$a = 64, \quad x_{32} = x_{16} = x_8 = x_4 = x_4 = x_2 = x_1 = y_{32} = y_{16}$$

$$= y_8 = y_4 = y_2 = y_1 = 2,$$

and let all other x_i, y_i be equal to 1. By $(*)$ and the above construction, we see that there exists a coloring of grid P satisfying condition (1), and there are two rows with 32 red squares each. Hence, $m \geqslant 32$.

Finally, we show $m \leqslant 32$. As some $x_i \geqslant 2$, $2m \leqslant a \leqslant 69$, $m \leqslant 34$. We show that $m = 33$ or 34 is impossible.

It only requires to consider $a \in \{66, 67, 68, 69\}$. For 64×64 table Q and any $r, s \leqslant a$, note that the number of red squares in any r rows but s columns is at most $r(a - s)$. We shall repeatedly use this property to derive contradictions.

If $a = 69$, then $x_m = 2$ and all other x_i's are 1. Evidently, $y_m = 2$ and all other y_i's are 1. In Q, the five rows with the most red squares have at least $69 + 68 + 67 + 66 + 66$ of them, while the five columns with the least red squares have $1 + 2 + 3 + 4 + 5$ of them, and

$$(69 + 68 + 67 + 66 + 66) - (1 + 2 + 3 + 4 + 5) > 5 \cdot (69 - 5)$$

leads to a contradiction.

If $a = 68$, as

$$(68 + 67 + 66 + 66) - (1 + 2 + 3 + 4) > 4 \cdot (68 - 4),$$

at least one of y_1, y_2, y_3, y_4 is 2 or more. As

$$(68 + 67 + 66 + 66 + 65 + 64 + 63 + 62)$$
$$-(1 + 2 + 3 + 4 + 5 + 6 + 7 + 8) - 8 \cdot (68 - 8) > 4,$$

we have $\sum_{i=1}^{8} (y_i - 1) \geqslant 2$. Further, as

$$(68 + 67 + 66 + 66 + 65 + \cdots + 54)$$
$$-(1 + 2 + \cdots + 16) - 16 \cdot (68 - 16) > 4 + 8,$$

we have $\sum_{i=1}^{16} (y_i - 1) \geqslant 3$, but this contradicts $\sum_{i=1}^{68} (y_i - 1) = 2$.

If $a = 67 (m = 33)$, as

$$(67 + 66 + 66) - (1 + 2 + 3) > 3 \cdot (67 - 3),$$

at least one of y_1, y_2, y_3 is 2 or more. As

$$(67 + 66 + 66 + 65 + 64 + 63) - (1 + 2 + 3 + 4 + 5 + 6) - 6 \cdot (67 - 6) > 3,$$

we have $\sum_{i=1}^{6} (y_i - 1) \geqslant 2$. As

$$(67 + 66 + 66 + 65 + \cdots + 57) - (1 + 2 + \cdots + 12) - 12 \cdot (67 - 12) > 3 + 6,$$

we have $\sum_{i=1}^{12} (y_i - 1) \geqslant 3$. Further, as

$$(67 + 66 + 66 + 65 + \cdots + 45)$$
$$-(1 + 2 + \cdots + 24) - 24 \cdot (67 - 24) > 3 + 6 + 12,$$

we have $\sum_{i=1}^{24} (y_i - 1) \geqslant 4$, but this contradicts $\sum_{i=1}^{67} (y_i - 1) = 3$.

If $a = 66 (m = 33)$, there are two $i x_i = 66$, and thus $i y_i \geq 2$ for all i, $y_1 \geq 2$. As

$$(66 + 66 + 65) - (1 + 2) - 3 \cdot (66 - 2) > 1,$$

we have $\sum_{i=1}^{2} (y_i - 1) \geq 2$. As

$$(66 + 66 + 65 + 64 + 63) - (1 + 2 + 3 + 4) - 5 \cdot (66 - 4) > 1 + 2,$$

we have $\sum_{i=1}^{4} (y_i - 1) \geq 3$. As

$$(66 + 66 + 65 + \cdots + 59) - (1 + 2 + \cdots + 8) - 9 \cdot (66 - 8) > 1 + 2 + 4,$$

we have $\sum_{i=1}^{8} (y_i - 1) \geq 4$. Further, as

$$(66 + 66 + 65 + \cdots + 51) - (1 + 2 + \cdots + 16) - 17 \cdot (66 - 16) > 1 + 2 + 4 + 8,$$

we have $\sum_{i=1}^{16} (y_i - 1) \geq 5$, but this contradicts $\sum_{i=1}^{66} (y_i - 1) = 4$.

To sum up, the maximum is $m = 32$.

Remark 1 The coloring is not unique. For example, let $a = 65$, $x_{32} = 2$, $y_{16} = y_8 = y_4 = 2, y_2 = 3$ and let all other x_i, y_i be 1. First, color the squares in row i, column j with $i + j \leq 65$; then color the first 32 squares in row 65 and the next 32 squares in row 66; finally, in columns 65 through 70, color 1, 2, 2, 4, 8, 16 squares respectively in rows 1 through 33. It can be checked that all conditions are satisfied.

Remark 2 In general, for $n \times n$ grid, it can be shown that $\sum_{i=1}^{2^k} (y_i - 1) \geq k$. The maximum m is given by $2m + \log_2 m \approx n$.

Test II, Second Day
(8 am–12:30 pm; 17 March 2023)

4. Call a nonempty set A of integers 'a beautiful set', if for any $a \in A$ and $k \in \{1, 2, \ldots, 2023\}$, the set

$$\left\{ b \in A : \left\lfloor \frac{b}{3^k} \right\rfloor = \left\lfloor \frac{a}{3^k} \right\rfloor \right\}$$

has exactly 2^k elements.

Prove that for a set S of integers, if the intersection of S and any beautiful set is nonempty, then S contains some beautiful set.

(Contributed by An Jinpeng)

Solution 1 For positive integer n, call a nonempty set A of integers 'a strictly beautiful set of order n', if $A \subset \{0, 1, \ldots, 3^n - 1\}$ and for any $a \in A$ and $1 \leqslant k \leqslant n$, there are 2^k elements in the set $\{b \in A \mid \lfloor 3^{-k}b \rfloor = \lfloor 3^{-k}a \rfloor\}$. Clearly, a strictly beautiful set of order 2023 is a beautiful set. We prove the following proposition by induction:

Proposition *For any positive integer n, if S is a set of integers and satisfies that the intersection of S and any strictly beautiful set of order n is nonempty, then S contains a strictly beautiful set of order n.*

Taking $n = 2023$, it is the problem statement.

If $n = 1$, a strictly beautiful set of order 1 is a 2-subset of $\{0, 1, 2\}$ and the proposition is true. Suppose the proposition is true when $n = m$; we show it is also true when $n = m + 1$. First, note that if A_1, A_2 are strictly beautiful sets of order m, and $\{i_1, i_2\}$ is a 2-subset of $\{0, 1, 2\}$, then

$$\{a + i_1 3^m \mid a \in A_1\} \cup \{a + i_2 3^m \mid a \in A_2\}$$

is a strictly beautiful set of order $m + 1$. Suppose the intersection of integer set S and any strictly beautiful set of order $m + 1$ is nonempty. Consider the sets

$$S_i = \{0 \leqslant a < 3^m \mid a + i 3^m \in S\}, \quad i = 0, 1, 2.$$

We assert that there exists a 2-subset $\{i_1, i_2\} \subset \{0, 1, 2\}$ such that the intersection of $S_{i_s}(s = 1, 2)$ and any strictly beautiful set of order m is nonempty. Otherwise, for some 2-subset $\{j_1, j_2\} \subset \{0, 1, 2\}$ and strictly beautiful sets B_1, B_2, we have $S_{j_s} \cap B_s = \varnothing (s = 1, 2)$, and then the strictly beautiful set

$$\{a + j_1 3^m \mid a \in B_1\} \cup \{a + j_2 3^m \mid a \in B_2\}$$

of order $m + 1$ does not intersect S, a contradiction. By the induction hypothesis, S_{i_1} contains a strictly beautiful set A_1 of order m, S_{i_2} contains a strictly beautiful set A_2 of order m as well. They together imply that S contains a strictly beautiful set of order $m + 1$, that is, $\{a + i_1 3^m \mid a \in A_1\} \cup \{a + i_2 3^m \mid a \in A_2\}$.

Solution 2 For integers m, n and $n > 0$, we call a nonempty set A of integers 'an m-beautiful set of order n', if $A \subset \{3^n m, 3^n m + 1, \ldots, 3^n(m + 1) - 1\}$ and for any $a \in A$ and $1 \leqslant k \leqslant n$, there are 2^k elements in the set $\{b \in A \mid \lfloor 3^{-k}b \rfloor = \lfloor 3^{-k}a \rfloor\}$. Evidently, for any integer m, every m-beautiful set of order 2023 is a beautiful set. We prove the following proposition by induction:

Proposition P_n *For any set S of integers and any integer m, there exists an m-beautiful set A of order n such that $A \subset S$ or $A \subset \overline{S}$ (here $\overline{S} = \mathbf{Z} \backslash S$ is the complement set of S).*

Note that once P_n, in particular P_{2023} is proved, then the set S or $\overline{S}$ contains a beautiful set A. However, $S \cap A \neq \varnothing$, and thus $A \subset S$, completing the proof.

If $n = 1$, by the pigeonhole principle, at least two elements $a, b \in \{3m, 3m+1, 3m+2\}$ both belong to S or both belong to $\overline{S}$. So, $A = \{a, b\}$ is an m-beautiful set of order 1, $A \subset S$ or $A \subset \overline{S}$, yielding proposition P_1. Suppose we have P_n and want to prove P_{n+1}. For any set S of integers and any integer m, by the induction hypothesis, there exist: a $3m$-beautiful set of order n,

$$A_1 \subset [3^n \cdot 3m, 3^n(3m+1)),$$

a $(3m+1)$-beautiful set of order n,

$$A_2 \subset [3^n(3m+1), 3^n(3m+2)),$$

and a $(3m+2)$-beautiful set of order n,

$$A_3 \subset [3^n(3m+2), 3^n(3m+3)).$$

They satisfy: for any $i \in \{1, 2, 3\}$, either $A_i \subset S$ or $A_i \subset \overline{S}$. By the pigeonhole principle, among A_1, A_2, A_3, at least two sets A_{i_1}, A_{i_2} $(1 \leqslant i_1 < i_2 \leqslant 3)$ are both contained in either S or $\overline{S}$. Define $A = A_{i_1} \cup A_{i_2}$. It is easy to check that A is an m-beautiful set of order $n+1$ and either $A \subset S$ or $A \subset \overline{S}$. This verifies P_{n+1} and the problem is solved.

5 Given positive integer n and n^3 integers $a_{ijk} \in \{1, -1\}(1 \leqslant i, j, k \leqslant n)$. Prove that there exist

$$x_1, \ldots, x_n, y_1, \ldots, y_n, z_1, \ldots, z_n \in \{1, -1\},$$

such that the following inequality holds:

$$\left| \sum_{i=1}^{n} \sum_{j=1}^{n} \sum_{k=1}^{n} a_{ijk} x_i y_j z_k \right| > \frac{n^2}{3}.$$

(Contributed by Deng Yu)

Solution 1　For any $(x_i), (y_j)$ taking values 1 or -1, we define

$$X_k = \sum_{i,j=1}^{n} a_{ijk} x_i y_j.$$

Taking $z_k = \operatorname{sgn} X_k$, we get

$$\sum_{i=1}^{n}\sum_{j=1}^{n}\sum_{k=1}^{n} a_{ijk} x_i y_j z_k = \sum_{k=1}^{n} |X_k|.$$

Sum over all possible $(x_i), (y_j)$ to obtain

$$\sum_{(x_i),(y_j)} |X_k|^2 = \sum_{(x_i),(y_i)} \sum_{i_1,i_2} \sum_{j_1,j_2} a_{i_1 j_1 k} a_{i_2 j_2 k} x_{i_1} x_{i_2} y_{j_1} y_{j_2}$$

$$= \sum_{i_1,i_2} \sum_{j_1,j_2} a_{i_1 j_1 k} a_{i_2 j_2 k} \sum_{(x_i),(y_j)} x_{i_1} x_{i_2} y_{j_1} y_{j_2}.$$

In the second line, the latter sum does not vanish only if $i_1 = i_2$ and $j_1 = j_2$. Hence,

$$\sum_{(x_i),(y_j)} |X_k|^2 = (2^n)^2 \sum_{i,j=1}^{n} a_{ijk}^2 = 2^{2n} n^2.$$

(To find a lower bound of $\sum_{(x_i),(y_j)} |X_k|$, we need to estimate $\sum_{(x_i),(y_j)} |X_k|^n$ for some $n > 2$.) Similarly,

$$\sum_{(x_i),(y_j)} |X_k|^4 = \sum_{i_1,i_2,i_3,i_4} \sum_{j_1,j_2,j_3,j_4} a_{i_1 j_1 k} a_{i_2 j_2 k} a_{i_3 j_3 k} a_{i_4 j_4 k}$$

$$\times \sum_{(x_i),(y_j)} x_{i_1} x_{i_2} x_{i_3} x_{i_4} y_{j_1} y_{j_2} y_{j_3} y_{j_4}.$$

The last sum does not vanish only if $(i_1, i_2, i_3, i_4), (j_1, j_2, j_3, j_4)$ can be made into identical (i, j) pairs (it is counted more than once when the four pairs are all identical). Hence,

$$\sum_{i_1,i_2,i_3,i_4} \sum_{j_1,j_2,j_3,j_4} a_{i_1 j_1 k} a_{i_2 j_2 k} a_{i_3 j_3 k} a_{i_4 j_4 k}$$

$$\times \sum_{(x_i),(y_j)} x_{i_1} x_{i_2} x_{i_3} x_{i_4} y_{j_1} y_{j_2} y_{j_3} y_{j_4} < 9n^4 (2^n)^2.$$

Now apply Hölder's inequality to derive

$$\left(\sum_{(x_i),(y_j)} \left(|X_k|^{\frac{2}{3}} \right)^{\frac{3}{2}} \right)^{\frac{2}{3}} \left(\sum_{(x_i),(y_j)} \left(|X_k|^{\frac{4}{3}} \right)^3 \right)^{\frac{1}{3}} \geqslant \sum_{(x_i),(y_j)} |X_k|^2 = 2^{2n} n^2,$$

and thus

$$\left(\sum_{(x_i),(y_j)} |X_k| \right)^{\frac{2}{3}} \cdot (2^{2n} \cdot 9n^4)^{\frac{1}{3}} > 2^{2n} n^2,$$

yielding

$$\sum_{(x_i),(y_j)} |X_k| > \left((2^{2n})^{\frac{2}{3}} \frac{n^{\frac{2}{3}}}{3^{\frac{2}{3}}} \right)^{\frac{3}{2}} = 2^{2n} \cdot \frac{n}{3}. \qquad \text{⑧}$$

Summing over k, it follows:

$$\sum_{(x_i),(y_j)} \sum_k |X_k| > 2^{2n} \cdot \frac{n^2}{3},$$

indicating that for some (x_i, y_j), $\sum_k |X_k| > \frac{n^2}{3}$.

Remark In the last part, instead of using Hölder's inequality, we can use the inequality $(X - 3n)^2 (X + 6n)X = X^4 - 27n^2 X^2 + 54n^3 X \geqslant 0$ when $X \geqslant 0$:

$$\sum_{(x_i),(y_j)} |X_k|^4 - 27n^2 \sum_{(x_i),(y_j)} |X_k|^2 + 54n^3 \sum_{(x_i),(y_j)} |X_k| \geqslant 0.$$

Therefore,

$$\sum_{(x_i),(y_j)} |X_k| \geqslant \frac{1}{54} \cdot 2^{2n} \cdot \frac{27n^2 \cdot n^2 - 9n^4}{n^3} = \frac{1}{3} \cdot 2^{2n} n.$$

Summing over k to obtain ⑧.

Solution 2 First, we need two lemmas.

Lemma 1 *Let n be a positive integer and $a_1, a_2, \ldots, a_n$ be real numbers. Then*

$$\sum_{x_1, \ldots, x_n \in \{-1, 1\}} \left| \sum_{i=1}^n a_i x_i \right| \geqslant 2 \binom{n-1}{\lfloor \frac{n-1}{2} \rfloor} \sum_{i=1}^n |a_i|.$$

Proof of Lemma 1 As every x_i traverses $\{-1, 1\}$, in the above inequality we may replace a_i by $|a_i|$, that is, assume $a_i \geqslant 0$ for $i = 1, 2, \ldots, n$. Note that when all x_i's change signs, the value of $\left| \sum_{i=1}^{n} a_i x_i \right|$ does not change. Thus, we have

$$\sum_{x_1, \ldots, x_n \in \{-1, 1\}} \left| \sum_{i=1}^{n} a_i x_i \right| \geqslant \sum_{\substack{x_1, \ldots, x_n \in \{-1, 1\} \\ x_1 + \cdots + x_n \neq 0}} \left| \sum_{i=1}^{n} a_i x_i \right|$$

$$= 2 \sum_{\substack{x_1, \ldots, x_n \in \{-1, 1\} \\ x_1 + \cdots + x_n > 0}} \left| \sum_{i=1}^{n} a_i x_i \right|$$

$$\geqslant 2 \sum_{\substack{x_1, \ldots, x_n \in \{-1, 1\} \\ x_1 + \cdots + x_n > 0}} \sum_{i=1}^{n} a_i x_i$$

$$= 2 \sum_{i=1}^{n} \left(\sum_{\substack{x_1, \ldots, x_n \in \{-1, 1\} \\ x_1 + \cdots + x_n > 0}} x_i \right) a_i.$$

By symmetry, observe that

$$\sum_{\substack{x_1, \ldots, x_n \in \{-1, 1\} \\ x_1 + \cdots + x_n > 0}} x_i$$

does not depend on i; in particular, for $i = n$, the value is

$$\sum_{\substack{x_1, \ldots, x_n \in \{-1, 1\} \\ x_1 + \cdots + x_n > 0}} x_i = \sum_{\substack{x_1, \ldots, x_{n-1} \in \{-1, 1\} \\ x_1 + \cdots + x_{n-1} > -1}} 1 + \sum_{\substack{x_1, \ldots, x_{n-1} \in \{-1, 1\} \\ x_1 + \cdots + x_{n-1} > 1}} (-1)$$

$$= \sum_{\substack{x_1, \ldots, x_{n-1} \in \{-1, 1\} \\ x_1 + \cdots + x_{n-1} \in \{0, 1\}}} 1$$

$$= \binom{n-1}{\left\lfloor \frac{n-1}{2} \right\rfloor}.$$

The last equality holds because $x_1, x_2, \ldots, x_{n-1} \in \{-1, 1\}$ satisfy $x_1 + x_2 + \cdots + x_{n-1} \in \{0, 1\}$ if and only if exactly $\left\lceil \frac{n-1}{2} \right\rceil$ of them take value 1 and the other $\left\lfloor \frac{n-1}{2} \right\rfloor$ of them take value -1. Now Lemma 1 is proved.

Lemma 2 *For any positive integer n,*

$$\binom{n-1}{\lfloor \frac{n-1}{2} \rfloor} \geqslant \frac{2^{n-1}}{\sqrt{2n}}.$$

Proof of Lemma 2 Evidently, the inequality holds when $n = 1, 2, 3$ (equal when $n = 2$). When $n = 2m \geqslant 4$, it is equivalent to $\binom{2m}{m} \geqslant \frac{2^{2m}}{\sqrt{4m}}$; when $n = 2m + 1 \geqslant 5$, it is equivalent to $\binom{2m}{m} \geqslant \frac{2^{2m}}{\sqrt{4m+2}}$. So, it suffices to prove, for any integer $m \geqslant 2$,

$$\binom{2m}{m} \geqslant \frac{2^{2m-1}}{\sqrt{m}}.$$

In fact,

$$\binom{2m}{m} = \prod_{k=1}^{m} \frac{2k(2k-1)}{k^2} = 2^{2m} \prod_{k=1}^{m} \frac{2k-1}{2k}.$$

Let

$$A = \prod_{k=2}^{m} \frac{2k-1}{2k}, \quad B = \prod_{k=2}^{m} \frac{2k-2}{2k-1}.$$

Then $A > B$ and $AB = \frac{2}{2m} = \frac{1}{m}$. Hence $A > \frac{1}{\sqrt{m}}$, yielding

$$\binom{2m}{m} = 2^{2m} \cdot \frac{1}{2} A > 2^{2m} \cdot \frac{1}{2\sqrt{m}} = \frac{2^{2m-1}}{\sqrt{m}}.$$

This verifies Lemma 2.

By the two lemmas, we immediately arrive at the following result: for any n real numbers $a_1, a_2, \ldots, a_n$,

$$\sum_{x_1,\ldots,x_n \in \{-1,1\}} \left| \sum_{i=1}^{n} a_i x_i \right| \geqslant \frac{2^n}{\sqrt{2n}} \sum_{i=1}^{n} |a_i|.$$

Using the result twice, we obtain, for any n^2 real numbers a_{ij},

$$\sum_{x_1,\ldots,x_n,y_1,\ldots,y_n \in \{-1,1\}} \left| \sum_{i=1}^{n} \sum_{j=1}^{n} a_{ij} x_i y_j \right|$$

$$= \sum_{x_1,\ldots,x_n \in \{-1,1\}} \left(\sum_{y_1,\ldots,y_n \in \{-1,1\}} \left| \sum_{j=1}^{n} \left(\sum_{i=1}^{n} a_{ij} x_i \right) y_j \right| \right)$$

$$\geqslant \sum_{x_1,\ldots,x_n,y_1,\ldots,y_n\in\{-1,1\}}\left(\frac{2^n}{\sqrt{2n}}\sum_{j=1}^{n}\left|\sum_{i=1}^{n}a_{ij}x_i\right|\right)$$

$$=\frac{2^n}{\sqrt{2n}}\sum_{j=1}^{n}\sum_{x_1,\ldots,x_n\in\{-1,1\}}\left|\sum_{i=1}^{n}a_{ij}x_i\right|$$

$$\geqslant\frac{2^n}{\sqrt{2n}}\sum_{j=1}^{n}\frac{2^n}{\sqrt{2n}}\sum_{i=1}^{n}|a_{ij}|$$

$$=\frac{2^{2n}}{2n}\sum_{i=1}^{n}\sum_{j=1}^{n}|a_{ij}|. \tag{9}$$

Return to the original problem. For fixed $x_1,\ldots,x_n,y_1,\ldots,y_n$, take $z_k\in\{-1,1\}$ such that

$$z_k\sum_{i=1}^{n}\sum_{j=1}^{n}a_{ijk}x_iy_j\geqslant 0,\quad k=1,2,\ldots,n.$$

It follows that

$$\left|\sum_{i=1}^{n}\sum_{j=1}^{n}\sum_{k=1}^{n}a_{ijk}x_iy_jz_k\right|=\sum_{k=1}^{n}\left|\sum_{i=1}^{n}\sum_{j=1}^{n}a_{ijk}x_iy_j\right|:=T_{x_1,\ldots,x_n,y_1,\ldots,y_n}.$$

From (9), we have

$$\sum_{x_1,\ldots,x_n,y_1,\ldots,y_n\in\{-1,1\}}T_{x_1,\ldots,x_n,y_1,\ldots,y_n}\geqslant\sum_{k=1}^{n}\frac{2^{2n}}{2n}\sum_{i=1}^{n}\sum_{j=1}^{n}|a_{ijk}|=2^{2n}\cdot\frac{n^2}{2}.$$

Since the average of the T's is no less than $\frac{n^2}{2}$, there must be some $x_1,\ldots,x_n,y_1,\ldots,y_n\in\{-1,1\}$ such that

$$T_{x_1,\ldots,x_n,y_1,\ldots,y_n}\geqslant\frac{n^2}{2},$$

as desired.

6 Prove that there exists a constant $\lambda>0$ such that for any real number $D>1$, we can find an acute triangle in the rectangular coordinate system that satisfies the following conditions:

(1) all three vertices are integer points (both coordinates are integers);

(2) all three side lengths are greater than D;

(3) the area is less than $\frac{\sqrt{3}}{4}D^2 + \lambda \cdot D^{\frac{4}{5}}$.

(Contributed by Zhang Sihui)

Solution We start with a lemma.

Lemma *Let $A = 10^6$. For any $0 < \epsilon < 1, n \in \mathbf{R}_+$, there exists $t \in \mathbf{Z} \cap [n - \frac{A}{\epsilon}, n]$, such that $\{\sqrt{3}t\} < \epsilon$. Here, $\{x\}$ is the decimal part of x.*

Proof of Lemma Apply induction to $[-\log \epsilon]$.

- When $[-\log \epsilon] = 0, \frac{1}{10} < \epsilon < 1$, as

$$\{11\sqrt{3}\} = \{\sqrt{19^2 + 2}\} = \frac{2}{\sqrt{19^2 + 2} + 19} \in \left(\frac{1}{20}, \frac{1}{19}\right),$$

for $n_0 = [n]$, there exists $0 \leqslant s \leqslant 19$ such that

$$\{n_0\sqrt{3} - s \cdot 11\sqrt{3}\} < \frac{1}{19} < \epsilon.$$

From $\frac{A}{\epsilon} > A > 11s$, we get $n_0 - 11s \in \mathbf{Z} \cap [n - \frac{A}{\epsilon}, n]$, and the lemma holds.

- Suppose the lemma holds when $[-\log \epsilon] = k$. When $[-\log \epsilon] = k + 1$, let $\epsilon' = 10\epsilon$. By the induction hypothesis, there exists $t' \in \mathbf{Z} \cap [n - \frac{A}{\epsilon'}, n]$ such that $\{\sqrt{3}t'\} < \epsilon'$. Now let $(1 + \sqrt{3})(2 + \sqrt{3})^m = x_m + y_m\sqrt{3}$ $(x_m, y_m \in \mathbf{Z}_+)$. Multiplying it by the conjugate, we find $x_m^2 - 3y_m^2 = -2$. Take the least m with $y_m > 10^{k+2}$. Then $10^{k+2} < x_m < 10^{k+3} - 1$ (because $x_m = 2x_{m-1} + 3y_{m-1}, y_m = x_{m-1} + 2y_{m-1}$, it follows $x_m < 2y_m, 2x_{m-1} < y_m < 4y_{m-1}$, and hence $x_m < 5x_{m-1} < \frac{5}{2}y_m < 10y_{m-1} \leqslant 10^{k+3}$), while

$$\{y_m\sqrt{3}\} = \frac{2}{\sqrt{x_m^2 + 2} + x_m} \in \left(\frac{1}{x_m + 1}, \frac{1}{x_m}\right) \subset (10^{-k-3}, 10^{-k-2}).$$

Note that $10^{-k-1} < \epsilon' < 10^{-k}$ and $\{t'\sqrt{3}\} < \epsilon'$. So, there exists a maximal nonnegative integer $s \leqslant 1000$ such that $s \cdot \{y_m\sqrt{3}\} < \{t'\sqrt{3}\}$. We have $\{t'\sqrt{3} - s \cdot y_m\sqrt{3}\} < 10^{-k-2} < \epsilon$, and moreover,

$$t' - sy_m > n - \frac{A}{10\epsilon} - 1000 \cdot 10^{k+3} \geqslant n - \frac{A}{10\epsilon} - \frac{10^5}{\epsilon} > n - \frac{A}{\epsilon},$$

which validates the lemma.

Return to the original problem. Take any acute triangle $O_1O_2O_3$ whose vertices are at integer points and whose side lengths are greater than A^{10}, and let its area be λ_1. Define $\lambda = \max\{\lambda_1, 10\sqrt{3}A^2\}$. If $D \leqslant A^{10}$, triangle $O_1O_2O_3$ meets the problem conditions.

In the following, assume $D > A^{10}$. Take $\delta = D^{-\frac{1}{6}}$ and $N = \left(\frac{D+\sqrt{2}\delta}{2}\right)^2$. Clearly, we have

$$\delta \cdot \sqrt{N} = \delta \cdot \frac{D + \sqrt{2}\delta}{2} > \frac{D^{\frac{4}{5}}}{2} > A,$$

i.e., $\frac{A}{\delta} < \sqrt{N}$.

- Choose $\epsilon = \delta, n = \sqrt{N}$ in the lemma: there exists $x \in \left[\sqrt{N} - \frac{A}{\delta}, \sqrt{N}\right] \cap \mathbf{Z}$, such that $\{x\sqrt{3}\} < \delta$;
- Choose $\epsilon = \delta, n = \frac{A}{\delta} + \sqrt{N - x^2}$ in the lemma: there exists $y \in \left[\sqrt{N - x^2}, \sqrt{N - x^2} + \frac{A}{\delta}\right]$ such that $\{y\sqrt{3}\} < \delta$.

For the above choice of x and y, $x^2 + y^2 \geqslant N$; also, since $x + y \leqslant 2\sqrt{N} + \frac{A}{\delta} < 3\sqrt{N}, x^2 + y^2 < 9N$,

$$x^2 + y^2 - N < \frac{2A}{\delta}\sqrt{N - x^2} + \frac{A^2}{\delta^2} < \frac{2A}{\delta}\sqrt{\frac{2A}{\delta}\sqrt{N} - \frac{A^2}{\delta^2}} + \frac{A^2}{\delta^2}$$

$$< \left(\frac{2A}{\delta}\right)^{\frac{3}{2}}\sqrt{\frac{D + \sqrt{2}\delta}{2}} + \frac{A^2}{\delta^2}.$$

Now we choose $X(0,0), Y(2x, 2y), Z(x - \sqrt{3}y, y + \sqrt{3}x)$ to be the vertices of an equilateral triangle.

Near Z, choose an integer point $Z'(u, v)$ such that

$$|u - (x - \sqrt{3}y)| < \delta, \quad |v - (y + \sqrt{3}x)| < \delta.$$

Then we have

$$|XY| = 2\sqrt{x^2 + y^2} > D,$$

$$|YZ'| \geqslant |YZ| - |ZZ'| > 2\sqrt{x^2 + y^2} - \sqrt{2}\delta \geqslant 2\sqrt{N} - \sqrt{2}\delta = D,$$

$$|Z'X| \geqslant |XZ| - |ZZ'| > 2\sqrt{x^2 + y^2} - \sqrt{2}\delta \geqslant 2\sqrt{N} - \sqrt{2}\delta = D.$$

On the other hand,

$$S_{\triangle XYZ'} = \frac{1}{2}|XY| \cdot h(Z', XY) \leqslant \frac{1}{2}|XY| \cdot (h(Z, XY) + |ZZ'|)$$

$$\leqslant \frac{1}{2} \cdot 2\sqrt{x^2 + y^2} \cdot (\sqrt{3}\sqrt{x^2 + y^2} + 2\delta)$$

$$= \sqrt{3}(x^2 + y^2) + 2\delta\sqrt{x^2 + y^2}$$

$$\leqslant \sqrt{3}N + \sqrt{3}\left(\frac{2A}{\delta}\right)^{\frac{3}{2}}\sqrt{\frac{D + \sqrt{2}\delta}{2}} + \frac{\sqrt{3}A^2}{\delta^2} + 2\delta\sqrt{x^2 + y^2}.$$

For the four terms in the upper bound, further estimates are given as follows:

$$\sqrt{3}N = \frac{\sqrt{3}}{4}D^2 + \frac{\sqrt{6}}{2}D\delta + \frac{\sqrt{3}}{2}\delta^2 < \frac{\sqrt{3}}{4}D^2$$
$$+ \frac{\sqrt{6}}{2}D^{\frac{4}{5}} + \sqrt{3},$$

$$\sqrt{3}\left(\frac{2A}{\delta}\right)^{\frac{3}{2}}\sqrt{\frac{D + \sqrt{2}\delta}{2}} < 4\sqrt{3}A^2 \cdot D^{\frac{3}{10}} \cdot 2D^{\frac{1}{2}} = 8\sqrt{3}A^2 \cdot D^{\frac{4}{5}},$$

$$\frac{\sqrt{3}A^2}{\delta^2} < \sqrt{3}A^2 \cdot D^{\frac{4}{5}},$$

$$2\delta\sqrt{x^2 + y^2} < 2D^{-\frac{1}{5}} \cdot 3\sqrt{N} = 3\left(D^{\frac{4}{5}} + D^{-\frac{2}{5}} \cdot \sqrt{2}\right)$$
$$< 3D^{\frac{4}{5}} + 3\sqrt{2}.$$

Altogether, it follows that

$$S_{\triangle XYZ'} < \left(\frac{\sqrt{3}}{4}D^2 + \frac{\sqrt{6}}{2}D^{\frac{4}{5}} + \sqrt{3}\right) + 8\sqrt{3}A^2 \cdot D^{\frac{4}{5}}$$
$$+ \sqrt{3}A^2 \cdot D^{\frac{4}{5}} + \left(3D^{\frac{4}{5}} + 3\sqrt{2}\right)$$
$$< 10\sqrt{3}A^2 \cdot D^{\frac{4}{5}} + \frac{\sqrt{3}}{4}D^2 \leqslant \frac{\sqrt{3}}{4}D^2 + \lambda \cdot D^{\frac{4}{5}},$$

completing the proof.

Test III, First Day
(8 am–12:30 pm; 25 March 2023)

1 Is there a positive irrational number x, such that there are at most finitely many positive integers n satisfying that

$$\{kx\} \geqslant \frac{1}{n+1}$$

for every $k \in \{1, \ldots, n\}$? Here, for positive real number y, $\{y\}$ is the decimal part of y.

(Contributed by An Jinpeng)

Solution Such irrational number x does not exist. We give three solutions as follows.

Solution 1 Let x be irrational. In the sequence $\{x\}, \{2x\}, \ldots, \{nx\}, \cdots$, since all terms are distinct and dense in the interval $(0, 1)$, there are infinitely many positive integers d such that

$$\{dx\} = \min\{\{ax\} \mid a = 1, 2, \ldots, d\}.$$

Arrange them into an increasing sequence $d_1 < d_2 < \cdots$. We claim for every $n = d_i - 1 \ (i \geqslant 2)$,

$$\min\{\{kx\} \mid k = 1, \ldots, n\} = \{d_{i-1}x\} \geqslant \frac{1}{d_i}.$$

Suppose not, then $d_i \cdot \{d_{i-1}x\} < 1$, which implies that

$$\{d_{i-1}d_ix\} = \{d_i \cdot \lfloor d_{i-1}x \rfloor + d_i \cdot \{d_{i-1}x\}\} = d_i \cdot \{d_{i-1}x\}.$$

On the other hand,

$$\{d_id_{i-1}x\} = \{d_{i-1}\lfloor d_ix \rfloor + d_{i-1}\{d_ix\}\} \leqslant d_{i-1} \cdot \{d_ix\}.$$

It follows that $d_{i-1} \cdot \{d_ix\} \geqslant d_i \cdot \{d_{i-1}x\}$. However, we have $d_{i-1} < d_i$ and $\{d_ix\} < \{d_{i-1}x\}$, a contradiction.

Therefore, $n = d_i - 1$ satisfies the problem condition and there are infinitely many of them. The required irrational number x does not exist.

Solution 2 If x is irrational and satisfies the problem condition, then there exists $n_0 \in \mathbf{Z}_+$ such that for every $n \geqslant n_0$, one can find $k \in \{1, \ldots, n\}$ with $\{kx\} < \frac{1}{n+1}$. As x is irrational, $\{kx\} \neq 0$ for all $k \in \mathbf{Z}_+$. For n_0, there are $k_0 \in \{1, \ldots, n_0\}$ and $l_0 \in \mathbf{Z}$ such that

$$0 < k_0x - l_0 < \frac{1}{n_0 + 1}.$$

Suppose $(k_0, l_0) = 1$; otherwise, replace them by $\frac{k_0}{\gcd(k_0, l_0)}, \frac{l_0}{\gcd(k_0, l_0)}$ respectively and the above inequalities still hold.

Let $n_1 = \left\lfloor \frac{1}{k_0x - l_0} \right\rfloor > n_0$. There are $k_1 \in \{1, \ldots, n_1\}$ and $l_1 \in \mathbf{Z}$ such that

$$0 < k_1x - l_1 < \frac{1}{n_1 + 1} < k_0x - l_0.$$

Likewise, suppose $(k_1, l_1) = 1$. As k_0 and l_0 are coprime, k_1 and l_1 are coprime, we have $\frac{l_0}{k_0} \neq \frac{l_1}{k_1}$, and hence

$$1 \leqslant |k_0l_1 - k_1l_0| = |k_1(k_0x - l_0) - k_0(k_1x - l_1)|$$

$$< \max\{k_1(k_0x - l_0), k_0(k_1x - l_1)\}$$

$$\times (\text{since } k_1(k_0x - l_0) > 0, k_0(k_1x - l_1) > 0)$$

$$\leqslant \max\left\{\frac{k_1}{n_1}, \frac{k_0}{n_1+1}\right\} \leqslant 1, \quad \left(\text{since}\frac{1}{k_0 x - l_0} \geqslant n_1, k_0 x - l_0 \leqslant \frac{1}{n_1}\right)$$

which is a contradiction.

Solution 3 For any irrational x, the decimal parts $\{kx\}, k \in \mathbf{Z}_+$ are distinct and dense in the interval $(0,1)$. So, there are infinitely many positive integers m that satisfy

$$\{mx\} < \{kx\}, \quad \forall\, k = 1, 2, \dots, m - 1.$$

For each such $m \geqslant 2$, let the minimum of $\{x\}, \{2x\}, \dots, \{(m-1)x\}$ be $\beta = \{rx\}$. Now, for $k = 1, 2, \dots, m - 1$, it is always $\{kx\} \geqslant \beta$, and so the open interval $(kx - \beta, kx)$ contains no integer.

In the coordinate plane, consider the parallelogram whose vertices are at $O(0,0), A(m, mx), B(m, mx - \beta)$ and $C(0, -\beta)$. We can see that no integer points lie inside $OABC$, and only three integer points lie on the boundary of $OABC$, which are $O, D(m, \lfloor mx \rfloor)$ and $E(r, rx - \beta) = (r, \lfloor rx \rfloor)$. Then $\triangle ODE$ has no interior or boundary integer points except the three vertices. By Pick's theorem, its area is $\frac{1}{2}$. This implies the area of $OABC \geqslant 2S_{\triangle ODE} = 1$. However, the area of $OABC$ is also equal to $m \times \beta = m \times \{rx\}$, so $m\beta \geqslant 1, \beta \geqslant \frac{1}{m}$, and for $k = 1, 2, \dots, m - 1$,

$$\{kx\} \geqslant \{rx\} \geqslant \frac{1}{m}.$$

This means $n = m - 1$ satisfies the problem condition, and there are infinitely many of them. Hence, the desired irrational number x does not exist.

2 For nonempty finite set B of real numbers and real number x, define

$$d_B(x) = \min_{b \in B} |x - b|.$$

(1) Given positive integer m. Find the smallest real number λ, such that for any positive integer n and any real numbers $x_1, x_2, \dots, x_n \in [0, 1]$, there is always a set B of m real numbers, such that

$$d_B(x_1) + d_B(x_2) + \cdots + d_B(x_n) \leqslant \lambda n.$$

(2) Given positive integer m and positive number ϵ. Prove that there exist a positive integer n and nonnegative real numbers $x_1, x_2, \dots, x_n$, such that for any set B of m real numbers,

$$d_B(x_1) + d_B(x_2) + \cdots + d_B(x_n) > (1 - \epsilon)(x_1 + x_2 + \cdots + x_n).$$

(Contributed by Wang Bin)

Solution   (1) Consider two m-element sets

$$B_1 = \left\{ \frac{0}{2m-1}, \frac{2}{2m-1}, \dots, \frac{2m-2}{2m-1} \right\},$$

$$B_2 = \left\{ \frac{1}{2m-1}, \frac{3}{2m-1}, \dots, \frac{2m-1}{2m-1} \right\}.$$

For any $x \in [0, 1]$, there exists $k \in \{0, 1, \dots, m-1\}$ such that $\frac{2k}{2m-1} \leqslant x \leqslant \frac{2k+1}{2m-1}$, or $k \in \{1, 2, \dots, m-1\}$ such that $\frac{2k-1}{2m-1} \leqslant x \leqslant \frac{2k}{2m-1}$. Either way, we have $d_{B_1}(x) + d_{B_2}(x) = \frac{1}{2m-1}$, and thus

$$\sum_{i=1}^{n} d_{B_1}(x_i) + \sum_{i=1}^{n} d_{B_2}(x_i) = \sum_{i=1}^{n} (d_{B_1}(x_i) + d_{B_2}(x_i)) = \frac{n}{2m-1}.$$

Taking either $B = B_1$ or $B = B_2$ such that $\sum_{i=1}^{n} d_B(x_i) \leqslant \frac{n}{4m-2}$, we see that $\lambda = \frac{1}{4m-2}$ satisfies the problem condition.

Next, we show $\lambda < \frac{1}{4m-2}$ is not desirable. Take $n = 2m$ and n-element set $X = \left\{ \frac{0}{n-1}, \frac{1}{n-1}, \dots, \frac{n-1}{n-1} \right\}$. For any set $B = \{b_1, b_2, \dots, b_m\}$ of m real numbers, we divide X into disjoint subsets as follows:

$$X = \bigcup_{k=1}^{m} X_k, \quad \text{where } X_k = \{x \in X : d_B(x) = |x - b_k|, k \text{ minimal}\},$$

in other words, X_k consists of elements of X that are closest to b_k (if x is equidistant from two or more b_k's, choose the minimal k). Note that each X_k is a contiguous block of X when the elements are arranged in increasing order. Hence,

$$\sum_{x \in X_k} |x - b_k| \geqslant \max_{x \in X_k} x - \min_{x \in X_k} x = \frac{1}{n-1} \times (|X_k| - 1).$$

Since $|X_1| + |X_2| + \dots + |X_m| = n$,

$$\sum_{i=1}^{n} d_B(x_i) \geqslant \sum_{k=1}^{m} \frac{1}{n-1} \times (|X_k| - 1) = \frac{1}{n-1} \times (n - m) = \frac{n}{4m-2}.$$

This indicates that $\lambda < \frac{1}{4m-2}$ is not desirable. Therefore, $\lambda = \frac{1}{4m-2}$.

(2) Choose positive integers L, t sufficiently large, such that $\frac{L-m}{L} \times \frac{t-2}{t} > 1 - \epsilon$. Take

$$n = 1 + t + \dots + t^{L-1} = \frac{t^L - 1}{t - 1},$$

and n real numbers $x_1, x_2, \dots, x_n$ as

$$t^{L-1} \ t's, t^{L-2} \ t^{2\prime}s, \dots, t^1 \ t^{L-1\prime}s, \quad \text{and} \quad 1 \ t^{L\prime}s(\text{the sum is } L \times t^L).$$

For any set $B = \{b_1, b_2, \dots, b_m\}$ of m real numbers, if for some $k \in \{1, 2, \dots, L\}$, $[2t^{k-1}, (2t-2)t^{k-1}] \cap B = \varnothing$, then $d_B(t^k) > (t-2)t^{k-1}$, and for those t^{L-k} of t^k, the sum of $d_B(x)$ is

$$d_B(t^k) \times t^{L-k} > (t-2)t^{k-1} \times t^{L-k} = (t-2)t^{L-1}.$$

Consider L disjoint intervals $[2t^{k-1}, (2t-2)t^{k-1}]$ $(k = 1, 2, \ldots, L)$, at least $(L-m)$ of which have an empty intersection with B. So, the above inequality holds for at least $(L-m)$ intervals, and we obtain

$$\sum_{i=1}^{n} d_B(x_i) > (L-m) \times (t-2)t^{L-1}$$

$$= \frac{L-m}{L} \times \frac{t-2}{t} \times \sum_{i=1}^{n} x_i > (1-\epsilon) \cdot \sum_{i=1}^{n} x_i.$$

(2) An alternative solution. Take positive integer $n > e^{\frac{2m}{\epsilon}}$ and $x_i = \frac{1}{i}, i = 1, 2, \ldots, n$, which satisfy

$$S = \sum_{i=1}^{n} x_i = 1 + \frac{1}{2} + \cdots + \frac{1}{n} > \ln n > \frac{2m}{\epsilon}.$$

Let m-element set $B = \{b_1, b_2, \ldots, b_m\}$. For each x_i, we focus on

$$x_i - d_B(x_i) = \max_{k=1,\ldots,m} \{x_i - |x_i - b_k|\} \leqslant \sum_{k=1}^{m} (x_i - |x_i - b_k|)_+,$$

where $y_+ = \max\{y, 0\}$ is the positive part of real number y. Then,

$$T = \sum_{i=1}^{n} (x_i - d_B(x_i)) \leqslant \sum_{i=1}^{n} \sum_{k=1}^{m} (x_i - |x_i - b_k|)_+$$

$$= \sum_{k=1}^{m} \sum_{i=1}^{n} (x_i - |x_i - b_k|)_+.$$

When $x_i \leqslant \frac{b_k}{2}$, we have $x_i - |x_i - b_k| \leqslant 0$, and so $(x_i - |x_i - b_k|)_+ = 0$; when $x_i > \frac{b_k}{2}$ (that is, $i < \frac{2}{b_k}$), we have $x_i - |x_i - b_k| \leqslant b_k$. Hence,

$$\sum_{i=1}^{n} (x_i - |x_i - b_k|)_+ = \sum_{i=1}^{\lfloor \frac{2}{b_k} \rfloor} (x_i - |x_i - b_k|)_+ \leqslant \sum_{i=1}^{\lfloor \frac{2}{b_k} \rfloor} b_k = \left\lfloor \frac{2}{b_k} \right\rfloor \cdot 2 \leqslant 2,$$

$$T = \sum_{i=1}^{n} (x_i - d_B(x_i)) \leqslant \sum_{k=1}^{m} \sum_{i=1}^{n} (x_i - |x_i - b_k|)_+ \leqslant 2m,$$

$$\sum_{i=1}^{n} d_B(x_i) = S - T \geqslant S - 2m \geqslant S - \epsilon S = (1-\epsilon)S.$$

3 For convex quadrilateral $ABCD$, an interior point P is called a '*balanced point*', if it satisfies the following conditions:

(a) P does not lie on the diagonals AC, BD;
(b) Extend AP, BP, CP, DP to intersect $ABCD$ at points A', B', C', D', respectively, then

$$AP \cdot PA' = BP \cdot PB' = CP \cdot PC' = DP \cdot PD'.$$

Find the maximum possible number of balanced points of $ABCD$.

(Contributed by Fu Yunhao)

Solution Let AC and BD meet at point O. Suppose balanced point P is an interior point of $\triangle AOB$, as shown in Fig. 7.5.

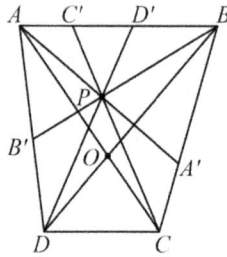

Fig. 7.5

From condition (b), we have

$$\angle PC'D' = 180° - \angle PC'A = 180° - \angle PA'C = \angle PA'B = \angle PB'A$$
$$= 180° - \angle PB'D = 180° - \angle PD'B = \angle PD'C' = \angle PCD.$$

Hence, $AB \parallel CD$. This indicates that if $ABCD$ is not a rectangle, a parallelogram or a trapezium, then it has no balanced points.

(1) Suppose $ABCD$ is a rectangle. If there is a balanced point in $\triangle AOB$ or $\triangle COD$, then P lies on the perpendicular bisector of AB, and $\angle APD = 90°$. This requires $AD > AB$, and there are at most two balanced points in $\triangle AOB$ and $\triangle COD$. Hence, for rectangle $ABCD$, there are at most two balanced points.

(2) Suppose $ABCD$ is a parallelogram. Let $\angle A > 90°$. We assert there is at most one balanced point in $\triangle AOB$ and it requires $AB < AD$. If this is

verified, then there are at most two balanced points in a nonrectangular parallelogram.

Suppose P is a balanced point in $\triangle AOB$. As illustrated in Fig. 7.6, we may assume that the perpendicular bisector of CD meets the segments AC, AB at interior points G, H, respectively (by property of P). Note that P is a balanced point if and only if $\angle APD = \angle C$. Let E be the symmetric point of A about H, $\angle EPC = \angle C = 180° - \angle B$, and so P lies on the circumcircle of $\triangle BCE$. Evidently, H is outside this circle. As E, G, D are collinear, we have G is inside this circle $\Leftrightarrow \angle EGC > \angle C \Leftrightarrow 2\angle GCD > \angle C \Leftrightarrow \angle ACD > \angle CAD \Leftrightarrow AD > CD$.

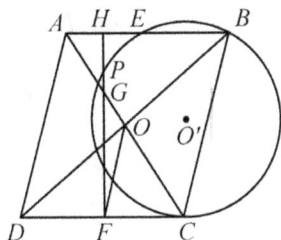

Fig. 7.6

Thus, when $AD > AB = CD$, there is exactly one balanced point.

When $AD \leqslant CD$, G is not inside the circumcircle of $\triangle BCE$. To have a balanced point inside $\triangle AOB$, we must require:

(a) if a perpendicular is dropped from the circumcenter O' of $\triangle BCE$ to GH, then the foot is inside GH;

(b) the distance from O' to GH does not exceed the radius $O'C$.

From (i), it follows that O' is between the lines AB and CD, $\angle B > 45°$.

From (ii), $O'C > GF$. Let X be the point on FH such that $OX \perp OF$. Then

$$FX = \frac{OF}{\cos \angle OFH} = \frac{BC}{2 \sin B} = O'C.$$

So, X cannot be an interior point of segment FG, $\angle FOG < 90°$, and $\angle DAC = \angle FOC > 90°$. However, if we let $CY \perp AD$ at Y, then

$CD \cdot \sin B = CY > DY > AD$, which implies that

$$O'C = \frac{BC}{2\sin B} < \frac{CD}{2},$$

and (ii) cannot hold. This contradiction shows that there are at most two balanced points in a nonrectangular parallelogram.

(3) Finally, suppose $ABCD$ is a non-parallelogram trapezium. If P is a balanced point in $\triangle AOB$, then $AB \parallel CD$. It follows that P lies on the perpendicular bisector of CD and $C'D'CD$ is an isosceles trapezium or a rectangle. Also, from the previous argument, we have $\angle PA'C = \angle PD'B$, and thus $\angle APD = \angle A'PD' = 180° - \angle B = \angle C$. If $ABCX$ is a parallelogram, then P lies on the circumcircle of $\triangle ADX$. Similarly, if Q is a balanced point in $\triangle COD$ and $BCDY$ is a parallelogram, then Q lies on the perpendicular bisector of AB and on the circumcircle of $\triangle ADY$. As a circle intersects a line at two points at most, there are at most two balanced points in $\triangle AOB$ and another two in $\triangle COD$. However, if both maxima are attained, then the circumcenter O_1 of $\triangle ADX$ is closer to AB than O, the circumcenter O_2 of $\triangle ADY$ is closer to CD than O (so O is closer to AB than O_2). It follows that O_1 is closer to AB than O_2, and the trapezium $ABCD$ must have two opposite acute angles and two opposite obtuse angles, as illustrated in Fig. 7.7 (E, F, G, H are as defined before).

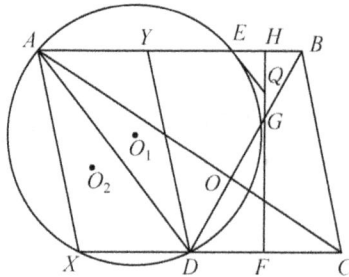

Fig. 7.7

Let the tangent of the circumcircle of $\triangle ADE$ through E intersect GH at point Q. Suppose that P_1, P_2 are the balanced points on GH: they must be inside segment QG, implying $QG > \sqrt{QP_1 \cdot QP_2} = QE$, and thus

$$\angle EGB = 2\angle EGQ < \angle EGQ + \angle GEQ = \angle EQH = \angle O_1EA$$

$$= 90° - \angle ADE = 90° - (\angle B - \angle A) < \angle A,$$

but G is outside the circumcircle of $\triangle ADE$, a contradiction. (Let J be the other point of intersection of BD and the circumcircle of $\triangle ADE$. We have $\angle EJB = \angle EAD > \angle EGB$, J is inside BG, and G must be inside the circumcircle of $\triangle ADE$.)

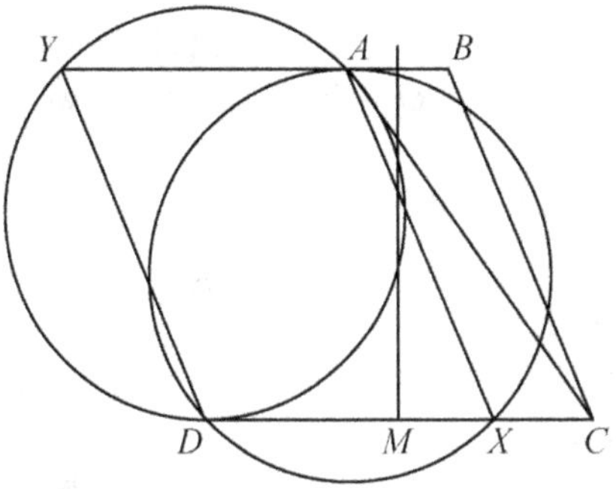

Fig. 7.8

Therefore, the number of balanced points in a trapezium cannot exceed 3. Now we have proved that the number of balanced points in an arbitrary convex quadrilateral is at most 3. For an isosceles trapezium, there can be three balanced points, as shown in Fig. 7.8. A specific example is: $A(-1,7), B(1,7), C(4,0), D(-4,0)$, and the three balanced points are at

$$\left(0, \frac{29 - \sqrt{57}}{7}\right), \quad \left(0, \frac{29 + \sqrt{57}}{7}\right), \quad \text{and} \quad \left(0, \frac{20 + 6\sqrt{22}}{7}\right).$$

In all, the maximum number of balanced points is 3.

Second solution for non-parallelogram trapezium We see that B' satisfies $\angle AB'B = \angle BB'C$, which is equivalent to $\frac{\sin \angle BCB'}{\sin A} = \frac{AB}{BC}$ by applying the sine rule in $\triangle ABB'$ and $\triangle BB'C$.

Suppose the rays DA and CB meet at point T, and $\angle C < 90°$. If P is a balanced point in $\triangle AOB$, then from $\angle BCB' < \angle C < 90°$ and

$$\sin \angle BCB' = \frac{AB}{BC} \sin A$$

is a fixed value, we see that B' is unique and hence P is unique. Similarly, if P is a balanced point in $\triangle COD$, then $\sin \angle D'AD$ is a fixed value, which correspond to two D', and two balanced points P in $\triangle COD$.

Third solution for non-parallelogram trapezium If P is a balanced point in $\triangle AOB$, then $AB \parallel CD$. It follows that P lies on the perpendicular bisector of CD and $C'D'CD$ is an isosceles trapezium or a rectangle.

Also, from the previous argument, we have $\angle PA'C = \angle PD'B$, and thus $\angle APD = \angle A'PD' = \pi - \angle B = \angle C$. Likewise, $\angle BPC = \angle B$. So, the balanced point P in $\triangle AOB$ is an intersection point of two circular arcs with chords AD and BC. There are at most two such points.

Suppose AB and CD are the only parallel sides in $ABCD$; P and Q are two balanced points in $\triangle AOB$. Then A, P, Q, D are concyclic, B, P, Q, C are concyclic. Assuming that Q is inside $\triangle CPD$, we have

$$\angle COD > \angle CQD = 360° - \angle PQC - \angle PQD$$

$$= \angle PBC + \angle PAD > \angle OBC + \angle OAD,$$

or $\angle COD - \angle OBC > \angle OAD$, that is, $\angle OCB > \angle OAD$, and thus the rays BC and AD intersect.

Under the above assumptions, there cannot be two balanced points in $\triangle COD$. Hence, there are at most three balanced points in the non-parallelogram trapezium $ABCD$.

Test III, Second Day
(8 am–12:30 pm; 26 March 2023)

4 Three mutually circumscribed circles $\odot O, \odot O_1, \odot O_2$ are located on the same side of line l, and they are tangent to l at points A, A_1, A_2, respectively, where A is on $A_1 A_2$. Let $\odot O$ and $\odot O_1$ touch at B_1, $\odot O$ and $\odot O_2$ touch at B_2, $\odot O_1$ and $\odot O_2$ touch at C. Suppose the lines $A_1 C$ and $A_2 B_2$ intersect at D_1, the lines $A_2 C$ and $A_1 B_1$ intersect at D_2. Prove that $D_1 D_2$ and l are parallel.

(Contributed by Yao Yijun)

We provide three solutions as follows.

Solution 1 Let P be the antipode of A in $\odot O$. Since in the right-angled trapezium $A_1 A O O_1$, $O_1 A_1 = O_1 B_1$, $OB_1 = OA$, we have

$$\angle A_1 B_1 A = 180° - \angle O_1 B_1 A_1 - \angle OB_1 A$$

$$= 180° - \frac{1}{2}(180° - \angle A_1 O_1 O) - \frac{1}{2}(180° - \angle AOO_1)$$

$$= \frac{1}{2}(\angle A_1 O_1 O + \angle AOO_1) = 90°,$$

and thus line A_1B_1 passes through point P; likewise, A_2B_2 passes through point P. Then,

$$\angle D_1B_2B_1 = \angle PB_2B_1 = \angle PAB_1 \quad (P, B_2, A, B_1 \text{ are concyclic})$$

$$= \angle B_1A_1A = \angle B_1CA_1 \quad (l \text{ is tangent to } \odot O_1)$$

$$= \angle B_1CD_1.$$

It follows that C, D_1, B_1, B_2 are concyclic; similarly, D_2 lies on the circumcircle of $\triangle CB_1B_2$. Therefore,

$$\angle D_1D_2B_1 = \angle D_1B_2B_1 = \angle B_1A_1A,$$

yielding $D_1D_2 \parallel l$.

Solution 2 We establish rectangular coordinates with line l as x axis, A as origin, and let $O(0,1)$. For $i = 1,2$, let r_i be the radius of $\odot O_i$. Since the coordinates of the centers $O_i(x_i, r_i)(i = 1, 2)$ satisfy

$$x_i^2 + (r_i - 1)^2 = (r_i + 1)^2,$$

we find $O_1(-2\sqrt{r_1}, r_1), O_2(2\sqrt{r_2}, r_2)$ and correspondingly, $A_1(-2\sqrt{r_1}, 0)$, $A_2(2\sqrt{r_2}, 0)$. Since $\odot O_1$ and $\odot O_2$ are circumscribed,

$$[2(\sqrt{r_1} + \sqrt{r_2})]^2 + (r_1 - r_2)^2 = (r_1 + r_2)^2,$$

which implies that

$$\sqrt{r_1} + \sqrt{r_2} = \sqrt{r_1 r_2}. \tag{$*$}$$

On the other hand, from

$$\overrightarrow{AB_1} = \frac{1 \cdot \overrightarrow{AO_1} + r_1 \cdot \overrightarrow{AO}}{r_1 + 1},$$

we find $B_1\left(-\frac{2\sqrt{r_1}}{r_1+1}, \frac{2r_1}{r_1+1}\right)$, and similarly, $B_2\left(\frac{2\sqrt{r_2}}{r_2+1}, \frac{2r_2}{r_2+1}\right)$, $C\left(\frac{2(r_1\sqrt{r_2}-r_2\sqrt{r_1})}{r_1+r_2}, \frac{2r_1r_2}{r_1+r_2}\right)$.

To show $D_1D_2 \parallel l$, it suffices to prove $y_{D_1} = y_{D_2}$. In $\triangle D_iA_1A_2$ ($i = 1, 2$), we have

$$y_{D_i} \cdot (\cot \angle D_iA_1A_2 + \cot \angle D_iA_2A_1) = A_1A_2.$$

Now we prove $\cot \angle D_1 A_1 A_2 + \cot \angle D_1 A_2 A_1 = \cot \angle D_2 A_1 A_2 + \cot \angle D_2 A_2 A_1$: the left and the right sides are given respectively by

$$\cot \angle D_1 A_1 A_2 + \cot \angle D_1 A_2 A_1 = \cot \angle C A_1 A_2 + \cot \angle B_2 A_2 A_1$$

$$= \frac{x_C - x_{A_1}}{y_C} + \frac{x_{A_2} - x_{B_2}}{y_{B_2}}$$

$$= \frac{\sqrt{r_1} + \sqrt{r_2}}{r_2} + \sqrt{r_2};$$

$$\cot \angle D_2 A_1 A_2 + \cot \angle D_2 A_2 A_1 = \cot \angle B_1 A_1 A_2 + \cot \angle C A_2 A_1$$

$$= \frac{x_{B_1} - x_{A_1}}{y_{B_1}} + \frac{x_{A_2} - x_C}{y_C}$$

$$= \sqrt{r_1} + \frac{\sqrt{r_1} + \sqrt{r_2}}{r_1}.$$

Taking the difference and using $(*)$, we find

$$\frac{\sqrt{r_1} + \sqrt{r_2}}{r_2} + \sqrt{r_2} - \sqrt{r_1} - \frac{\sqrt{r_1} + \sqrt{r_2}}{r_1}$$

$$= \sqrt{r_1 r_2} \left(\frac{1}{r_2} - \frac{1}{r_1} \right) + \sqrt{r_2} - \sqrt{r_1} = \frac{r_1 - r_2}{\sqrt{r_1 r_2}} + \sqrt{r_2} - \sqrt{r_1}$$

$$= (\sqrt{r_1} - \sqrt{r_2}) \left(\frac{\sqrt{r_1} + \sqrt{r_2}}{\sqrt{r_1 r_2}} - 1 \right) = 0.$$

So, $y_{D_1} = y_{D_2}$, and $D_1 D_2 \parallel l$.

Solution 3   In fact, the tangency of $\odot O$ and line l is not needed.

As shown in Fig. 7.9, we find: O, B_1, O_1 are collinear; O, B_2, O_2 are collinear; O_1, C, O_2 are collinear. Also, $OB_1 = OB_2, O_1 A_1 = O_1 B_1,$

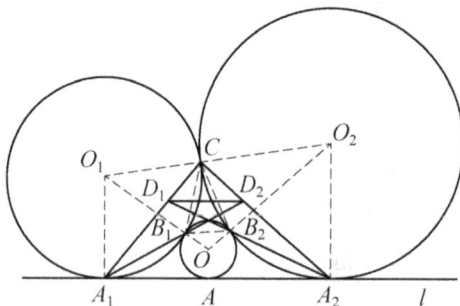

Fig. 7.9

$O_2A_2 = O_2B_2$, and O_1A_1, O_2A_2 are perpendicular to l. Moreover,

$$\angle CD_1B_2 = \angle CA_1A_2 + \angle A_1A_2B_2 = \angle CA_1B_1 + \angle B_1A_1A_2 + \angle A_1A_2B_2$$

$$= \frac{1}{2}\angle CO_1B_1 + \frac{1}{2}\angle A_1O_1B_1 + \frac{1}{2}\angle A_2O_2B_2$$

$$= (90° - \angle O_1B_1C) + \frac{1}{2}\angle O_1OO_2 \quad (\text{since } O_1A_1 \parallel O_2A_2)$$

$$= (\angle OB_1C - 90°) + 90° - \angle OB_1B_2$$

$$= \angle CB_1B_2.$$

Hence, C, D_1, B_1, B_2 are concyclic, and likewise C, D_2, B_2, B_1 are concyclic. We infer that C, D_1, B_1, B_2, D_2 are concyclic, $\angle CD_1D_2 = \angle CB_1D_2 = 180° - \angle A_1B_1C = \angle CA_1A_2$, yielding $D_1D_2 \parallel l$.

⑤ Is there a sequence of distinct integers $a_1, a_2, \ldots$, satisfying the two conditions

(a) For every positive integer k, $a_{k^2} > 0$ and $a_{k^2+k} < 0$;
(b) For every positive integer n, $|a_{n+1} - a_n| \leqslant 2023\sqrt{n}$?

(Contributed by Xiao Liang)

Solution There is no such sequence. We use proof by contradiction and assume it exists. Take positive integer N such that

$$\frac{1}{N+1} + \frac{1}{N+2} + \cdots + \frac{1}{N^2} > 2024.$$

The existence of N is due to

$$\sum_{k=N+1}^{N^2} \frac{1}{k} \geqslant \int_{N+1}^{N^2+1} \frac{1}{k}dk = \ln\frac{N^2+1}{N+1} > \ln(N-1)$$

can be arbitrarily large. Alternatively, we can choose $N = 2^m, m = 4048$:

$$\frac{1}{N+1} + \frac{1}{N+2} + \cdots + \frac{1}{N^2}$$

$$= \left(\frac{1}{2^m+1} + \cdots + \frac{1}{2^{m+1}}\right) + \left(\frac{1}{2^{m+1}+1} + \cdots + \frac{1}{2^{m+2}}\right)$$

$$+ \cdots + \left(\frac{1}{2^{2m-1}+1} + \cdots + \frac{1}{2^{2m}}\right)$$

$$> \frac{1}{2} + \frac{1}{2} + \cdots + \frac{1}{2} = \frac{m}{2} = 2024.$$

We assert that at least $4046N^2 + 2$ elements of the sequence $a_1, a_2, \ldots$ are in the interval $S = [-2023N^2, 2023N^2]$ and this will lead to a contradiction, as a_i's are required to be distinct. We shall prove, for $k = N, N+1, \ldots, N^2 - 1$,

(i) At least $\left\lfloor \frac{N^2}{k+1} \right\rfloor$ elements of $a_{k^2}, a_{k^2+1}, \ldots, a_{k^2+k-1}$ belong to S;

(ii) At least $\left\lfloor \frac{N^2}{k+1} \right\rfloor$ elements of $a_{k^2+k}, a_{k^2+k+1}, \ldots, a_{k^2+2k}$ belong to S.

Then, the size of S is greater than or equal to

$$2 \cdot \sum_{k=N+1}^{N^2} \left\lfloor \frac{N^2}{k} \right\rfloor \geqslant 2N^2 \sum_{k=N+1}^{N^2} \frac{1}{k} - 2(N^2 - N)$$

$$> 2N^2 \cdot 2024 - 2N^2 + 2N > 4046N^2 + 2.$$

We prove (i) below. Note for $l \in \{k^2, k^2 + 1, \ldots, k^2 + k - 1\}$,

$$|a_{l+1} - a_l| \leqslant 2023(k+1). \tag{$*$}$$

For the sequence $a_{k^2}, a_{k^2+1}, \ldots, a_{k^2+k-1}$, there are three situations.

(a) If some $a_l \geqslant 2023N^2$. Then $(*)$ implies that at least $\left\lfloor \frac{2023N^2}{2023(k+1)} \right\rfloor$ elements of $a_l, a_{l+1}, \ldots, a_{k^2+k-1}$ are in the interval $[0, 2023N^2]$;

(b) If some $a_l \leqslant -2023N^2$. Then $(*)$ implies that at least $\left\lfloor \frac{2023N^2}{2023(k+1)} \right\rfloor$ elements of $a_{k^2}, a_{k^2+1}, \ldots, a_l$ are in the interval $[-2023N^2, 0]$;

(c) If neither (a) nor (b) occurs. Then the sequence $a_{k^2}, a_{k^2+1}, \ldots, a_{k^2+k-1}$ is entirely in S, for a total of $k \geqslant \left\lfloor \frac{N^2}{k+1} \right\rfloor$ elements.

This proves (i); the proof of (ii) is similar and omitted. The contradiction indicates that such sequence does not exist.

6 Find the largest real number λ with the following property: for any doubly stochastic matrix of order 100, one can always keep 150 entries and change the other 9850 entries to zero such that in the resulting matrix, the sum of entries in every row and in every column is at least λ.

Note: A doubly stochastic matrix of order n is an $n \times n$ matrix in which all entries are nonnegative and the sum of entries in every row and in every column is equal to 1.

(Contributed by Qu Zhenhua)

Solution $\lambda = \frac{17}{1900}$.

First, we construct doubly stochastic matrix M as follows. Divide M into four submatrices: the first 75 rows, the first 24 columns form submatrix A; the last 25 rows, the first 24 columns form submatrix B; the first 75 rows, the last 76 columns form submatrix C; the last 25 rows, the last 76 columns form submatrix D (see Fig. 7.10). In A, all entries are $\frac{1}{75}$; in B, all entries are 0; in C, all entries are $\frac{17}{1900}$; in D, all entries are $\frac{1}{76}$. It is easy to verify M is a doubly stochastic matrix.

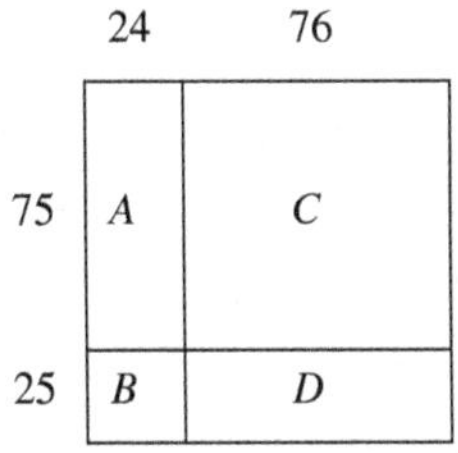

Fig. 7.10

Suppose we can choose 150 entries of M such that the sum in every row and column is greater than $\frac{17}{1900}$, and a elements are from A, d elements are from D. In the first 75 rows and last 76 columns, at least $151 - a - d$ rows or columns have no entries chosen from A or D, and each of them must contain at least 2 entries chosen from C. However, each entry chosen from C contributes to only one row and one column, implying that $2 \cdot (151 - a - d) \cdot \frac{1}{2} = 151 - a - d$ entries must be chosen from C, but then we have chosen $a + d + (151 - a - d) = 151$ entries from A, C and D, a contradiction. Hence, $\lambda \leqslant \frac{17}{1900}$.

Next, we show $\lambda = \frac{17}{1900}$ has the desired property. Let M be any doubly stochastic matrix of order 100. Call an entry *large* if it is greater than or equal to $\frac{17}{1900}$.

If there are 50 large entries in distinct rows and distinct columns. By interchanging rows and columns, we can assume they are in the first 50 rows and 50 columns. Choose the largest entry from each of the last 50 rows and last 50 columns, each entry is at least $\frac{1}{100} > \frac{17}{1900}$, the total number of entries chosen does not exceed 150, and the sum in each row or column is at least $\frac{17}{1900}$.

We claim the above situation always happens. Suppose to the contrary that no 50 large entries are in distinct rows and distinct columns. By Hall's theorem with deficiency,

There exist k rows ($51 \leqslant k \leqslant 100$), such that the large entries in these k rows occupy at most $k - 51$ columns. $\hspace{2em}$ (∗)

Suppose the large entries in the first k rows occupy the first $k - 51$ columns. Divide M into four submatrices A, B, C, D in the same order as in Fig. 7.10, where A has k rows and $k - 51$ columns. Let $S(A), S(B), S(C), S(D)$ be the sums of all entries in A, B, C, D, respectively. As the entries in the first $k - 51$ columns add up to $k - 51$, we have $S(A) + S(B) = k - 51$; there is no large entry in C, thus $S(C) < \frac{17}{1900} k(151 - k)$; in addition, $S(D) \leqslant S(B) + S(D) = 100 - k$, hence

$$100 = S(A) + S(B) + S(C) + S(D)$$

$$< (k - 51) + \frac{17}{1900} k(151 - k) + (100 - k)$$

$$\leqslant 49 + \frac{17}{1900} \cdot 75 \cdot 76 = 100,$$

which is a contradiction.

Therefore, the largest value of λ is $\frac{17}{1900}$.

Remark 1 We provide further explanation of (∗). For a bipartite graph $G = (U + V; E)$ and $A \subset U$, define the deficiency of G as

$$\operatorname{def}(A) = |A| - |N_G(A)|, \quad \delta = \max_{A \subset U} \operatorname{def}(A),$$

where $N_G(A)$ is the set of neighboring vertices of A. When $A = \varnothing$, define $\operatorname{def}(\varnothing) = 0$. So $\delta \geqslant 0$.

Hall's theorem with deficiency If the deficiency of a bipartite graph $G = (U + V; E)$ is δ, then there exists a matching with at least $|U| - \delta$ pairs. That is, at most δ vertices of U are unmatched.

Proof Adding δ dummy vertices $v_1, v_2, \ldots, v_\delta$ to V and connecting them to every vertex of U, we obtain a new bipartite graph $G^* = (U + V^*; E^*), V^* = V \cup \{v_1, v_2, \ldots, v_\delta\}$. The deficiency of G^* is $\delta(G^*) \geqslant 0$. Now, by the usual Hall's theorem, there exists a perfect matching in G^* (with $|U|$ pairs). Removing the dummy vertices, the matching includes at least $|U| - \delta$ pairs.

In the original problem, the locations of large entries give a bipartite graph $G = (U + V; E)$: let U be the 100 rows and V be the 100 columns; $u_i v_j$ is an edge if and only if the entry in row i and column j is large. As no 50 large entries are in distinct rows and distinct columns, there is no matching with 50 pairs. Hence, $\delta(G) \geqslant 51$, which is (∗).

Remark 2 We can also apply König-Egerváry theorem to prove there are 50 large entries in distinct rows and distinct columns. Suppose not, then by König-Egerváry theorem all large entries can be covered by 49 rows or columns, say the first l rows and the last 49-l columns. Divide the matrix into four submatrices similar to Fig. 7.10, where all large entries are in A, B or D. Let $S(A), S(B), S(C), S(D)$ be the sums of all entries in A, B, C, D, respectively. Then

$$S(A) + S(B) = l, \quad S(B) + S(D) = 49 - l.$$

Since C does not have large entries, $S(C) < \frac{17}{1900}(100 - l)(51 + l)$, and we find

$$100 = S(A) + S(B) + S(C) + S(D)$$
$$< l + (49 - l) + \frac{17}{1900}(100 - l)(51 + l)$$
$$\leqslant 49 + \frac{17}{1900} \cdot 75 \cdot 76 = 100,$$

a contradiction!

Test IV, First Day
(8 am–12:30 pm; 29 March 2023)

1. Let A, B be fixed points on the unit circle ω such that $\sqrt{2} < AB < 2$. Let P be a moving point on ω such that $\triangle ABP$ is an acute-angled triangle and $AP > AB > BP$. Let H be the orthocenter of $\triangle ABP$, S be a point on the minor arc $\overset{\frown}{AP}$ satisfying $SH = AH$, and T be a point on the minor arc $\overset{\frown}{AB}$ satisfying $TB \parallel AP$. Let Q be the intersection of lines ST and BP.

Prove that as P varies, the circle with diameter HQ passes through a fixed point on the plane.

(Contributed by He Yijie)

Solution 1 We claim that the circle with diameter HQ always passes through the midpoint M of AB.

As illustrated in Fig. 7.11, let P_1 be the antipode of P on ω and H_1 be the intersection of ω and the extension of AH. We show $QP_1 = QH_1$ as follows.

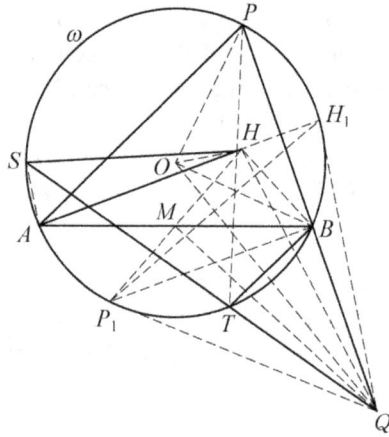

Fig. 7.11

Let O be the center of ω. Since $SH = AH$, S, A are symmetric about OH, $OH \perp SA$. Also, $HB \perp AP$, and thus $\angle BHO = 180° - \angle SAP = 180° - \angle STP = \angle PTQ$. Since $TB \parallel AP$, we find $\angle TPQ = \angle APB - \angle APT = \angle APB - \angle PAB = \angle HBO$. Hence $\triangle PTQ$ and $\triangle BHO$ are similar, $\frac{PQ}{PT} = \frac{BO}{BH}$. As $PT = AB, BO = PO$, it follows $\frac{PQ}{AB} = \frac{PO}{BH}$. Together with $\angle OPQ = 90° - \angle PAB = \angle ABH$, we find $\triangle OPQ$ and $\triangle HBA$ are similar, and hence $\angle OQP = \angle HAB = \angle H_1AB = \angle H_1P_1B$. As $PQ \perp P_1B$, so $OQ \perp P_1H_1$, OQ bisects P_1H_1, and we get $QP_1 = QH_1$.

Evidently, H_1, H are symmetric about BP, hence $QH = QH_1 = QP_1$; M is the midpoint of HP_1, hence $QM \perp MH$. This verifies that the circle with diameter HQ always passes through the midpoint M of AB.

Solution 2 Consider the complex plane with origin O. For point X, we use x to represent the complex number associated with it. First,

$$Q \in PB \Leftrightarrow \frac{q-p}{q-b} \in \mathbf{R} \Leftrightarrow \frac{q-p}{q-b} = \frac{\bar{q}-\bar{p}}{\bar{q}-\bar{b}}$$

$$\Leftrightarrow q\bar{q} - p\bar{q} - q\bar{b} + p\bar{b} = q\bar{q} - b\bar{q} - q\bar{p} + b\bar{p}$$

$$\Leftrightarrow (p-b)\bar{q} + \frac{p-b}{pb}q = \frac{p^2-b^2}{pb}$$

$$\Leftrightarrow q + pb\bar{q} = p + b.$$

Similarly,

$$Q \in ST \Leftrightarrow q + st\bar{q} = s + t.$$

Putting them together, we find

$$\bar{q} = \frac{s + t - p - b}{st - pb}.$$

As $BT \parallel AP$, $t = \frac{ap}{b}$; as $SH = AH$, s satisfies

$$(s - h)\left(\frac{1}{s} - \bar{h}\right) = (a - h)\left(\frac{1}{a} - \bar{h}\right)$$

$$\Leftrightarrow 1 - \frac{h}{s} - \bar{h}s + h\bar{h} = 1 - \frac{h}{a} - \bar{h}a + h\bar{h}$$

$$\Leftrightarrow \bar{h}s^2 - \left(\frac{h}{a} + \bar{h}a\right)s + h = 0.$$

Since $s = a$ is a root of the quadratic equation, by Vieta's formulas,

$$s = \frac{h}{a\bar{h}} = bp\frac{a + b + p}{ab + bp + pa}. \quad (h = a + b + p, |a| = |b| = |p| = 1)$$

Now we have

$$s + t - p - b = p \cdot \left[\frac{a}{b} + \frac{b(a + b + p)}{ab + bp + pa} - 1\right] - b$$

$$= p \cdot \frac{a(ab + bp + pa) + b(b^2 - pa)}{b(ab + bp + pa)} - b$$

$$= \frac{ap(ab + bp + pa) + bp(b^2 - pa) - b^2(ab + bp + pa)}{b(ab + bp + pa)}$$

$$= \frac{(ab + pa)(ap - b^2)}{b(ab + bp + pa)} = \frac{a(b + p)(ap - b^2)}{b(ab + bp + pa)};$$

$$st - pb = \frac{ap^2(a + b + p) - pb(ab + bp + pa)}{ab + bp + pa}$$

$$= \frac{p[ap(a + p) - b^2(a + p)]}{ab + bp + pa}$$

$$= \frac{p(ap - b^2)(a + p)}{ab + bp + pa}.$$

Therefore,

$$\bar{q} = \frac{a(p + b)}{bp(p + a)}, \quad q = \frac{\frac{1}{abp}(p + b)}{\frac{1}{abp^2}(p + a)} = \frac{p(p + b)}{p + a}.$$

As $h = a + b + p$, the circle with diameter HQ is centered at

$$\frac{1}{2}(h + q) = \frac{a + b}{2} + \frac{p}{2}\frac{(p + a) + (p + b)}{p + a},$$

with radius

$$\left|\frac{1}{2}(h - q)\right| = \left|\frac{1}{2(p + a)}[(a + b + p)(p + a) - p(p + b)]\right|$$

$$= \left|\frac{a}{2}\frac{(p + a) + (p + b)}{p + a}\right|.$$

From $|a| = |p| = 1$, it follows that the circle passes through the midpoint $\frac{a+b}{2}$ of AB.

2 Given integers a, b, and d, satisfying $d \geqslant 0, |a| \geqslant 2$ and $b \geqslant (|a| + 1)^{d+1}$. Let $f(x)$ be a polynomial of real coefficients of order d; for each positive integer n, let r_n be the remainder of $\lfloor f(n)a^n \rfloor$ modulo b.

Prove that if the sequence $\{r_n\}$ is eventually periodic, then $f(x)$ is a polynomial of rational coefficients.

Note 1: for $x \in \mathbf{R}$, $\lfloor x \rfloor$ is the largest integer not exceeding x.

Note 2: a sequence $\{r_n\}$ is eventually periodic, if there exist positive integers n_0, T, such that for any $n \geqslant n_0$, $a_{n+T} = a_n$.

(Contributed by Li Ting)

Solution   First, we give a lemma.

Lemma   *Let integer a and real number z satisfy $|a| \geqslant 2$ and for any nonnegative integer n,*

$$\|a^n z\| < \frac{1}{|a| + 1},$$

then z is an integer. Here, $\|x\|$ is the distance between x and the nearest integer,

$$\|x\| = \min(\{x\}, 1 - \{x\}).$$

Proof of Lemma   Assume $a \geqslant 2$ (otherwise replace a by $-a$). Suppose the conclusion is false. Assume $0 < z \leqslant \frac{1}{2}$ (otherwise replace z by $m \mp z$, where m is an integer).

As $z \in \left(0, \frac{1}{a^{-1}(a+1)}\right)$, we can assume

$$\frac{1}{a^k(a+1)} \leqslant z < \frac{1}{a^{k-1}(a+1)} \quad (k \in \mathbf{Z}_{\geqslant 0}),$$

which leads to $\frac{1}{a+1} \leqslant a^k z < 1 - \frac{1}{a+1}$, $\|a^k z\| \geqslant \frac{1}{a+1}$, a contradiction. The lemma is verified.

Back to the original problem. Suppose the statement is false and let $f(x) = a_d x^d + a_{d-1} x^{d-1} + \cdots + a_1 x + a_0$ have irrational coefficients $a_t, \ldots$, here t is the largest subscript for which $a_t \notin \mathbf{Q}$. Let

$$\left\{ \frac{f(n)a^n}{b} \right\} = \frac{r_n + \epsilon_n}{b}, \quad \epsilon_n \in [0, 1).$$

Define

$$g(x) = \Delta^t f(x) = \sum_{i=0}^{t} (-1)^i \binom{t}{i} f(x + t - i).$$

Then, except the constant term, all other terms of $g(x)$ have rational coefficients. Let $g(x) = c_{d-t} x^{d-t} + \cdots + c_1 x + c_0$ and K be the least common multiple of the denominators of $c_1, c_2, \ldots, c_{d-t}$. We have

$$\left\{ \frac{g(n)a^{n+t}}{b} \right\} = \left\{ \sum_{i=0}^{t} \frac{(-1)^i \binom{t}{i} f(n+t-i)a^{n+t-i} \cdot a^i}{b} \right\}$$

$$= \left\{ \sum_{i=0}^{t} \frac{(-1)^i \binom{t}{i} a^i (r_{n+t-i} + \epsilon_{n+t-i})}{b} \right\}$$

$$= \left\{ \sum_{i=0}^{t} r_{n+t-i} \frac{(-1)^i \binom{t}{i} a^i}{b} + \sum_{i=0}^{t} \epsilon_{n+t-i} \frac{(-1)^i \binom{t}{i} a^i}{b} \right\}.$$

Let $\delta_n = \sum_{i=0}^{t} \epsilon_{n+t-i} \cdot \frac{(-1)^i \binom{t}{i} a^i}{b}$. Note that

$$\sum_{i=0}^{t} \left| \frac{(-1)^i \binom{t}{i} a^i}{b} \right| = \sum_{i=0}^{t} \frac{\binom{t}{i} |a|^i}{b} = \frac{(|a|+1)^t}{b} \leqslant \frac{(|a|+1)^d}{b} \leqslant \frac{1}{|a|+1}.$$

It follows for any positive integers n, n',

$$|\delta_n - \delta_{n'}| \leqslant \sum_{i=0}^{t} |\epsilon_{n+t-i} - \epsilon_{n'+t-i}| \cdot \frac{\binom{t}{i} a^i}{b} < \sum_{i=0}^{t} \frac{\binom{t}{i} |a|^i}{b} \leqslant \frac{1}{|a|+1}.$$

We choose positive integers M and n, such that M is a multiple of the smallest period of $\{r_n\}$, $\varphi(bK)|M$, and for $n \geqslant n_0, bK|(a^{n+M+t} - a^{n+t})$. Now for $n \geqslant n_0$, we have

$$\left\| \frac{g(n+M)a^{n+M+t}}{b} - \frac{g(n)a^{n+t}}{b} \right\| = \|\delta_{n+M} - \delta_n\| < \frac{1}{|a|+1},$$

in which the equality is from $r_{n+M+t-i} = r_{n+t-i}, i = 0, 1, \ldots, t$. On the other hand,

$$\left\| \frac{g(n+M)a^{n+M+t}}{b} - \frac{g(n)a^{n+t}}{b} \right\|$$

$$= \left\| \frac{c_0}{b}(a^{n+M+t} - a^{n+t}) + \sum_{i=1}^{d-t} \frac{c_i}{b}\left((n+M)^i a^{n+M+t} - n^i a^{n+t}\right) \right\|$$

$$= \left\| \frac{c_0(a^M - 1)a^{n_0+t}}{b} \cdot a^{n-n_0} \right\|.$$

According to the lemma, $\frac{c_0(a^M-1)a^{n_0+t}}{b} \in \mathbf{Z}$, but c_0 is irrational. This invalidates the assumption and completes the proof.

3 Let $n \geqslant 2$ be a given integer. Find the minimum value of λ, such that for any real numbers $a_1, a_2, \ldots, a_n$ and b,

$$\lambda \sum_{i=1}^{n} \sqrt{|a_i - b|} + \sqrt{n \left| \sum_{i=1}^{n} a_i \right|} \geqslant \sum_{i=1}^{n} \sqrt{|a_i|}.$$

(Contributed by Li Ting)

Solution We claim that $\lambda_n = \frac{n-1+\sqrt{n-1}}{\sqrt{n}}$ is the desired smallest value. We start with a few lemmas.

Lemma 1 *The sequence $\{\lambda_n\}$ is strictly increasing.*

Proof of Lemma 1 Write

$$\lambda_n = \sqrt{n} - \frac{1}{\sqrt{n}} + \sqrt{1 - \frac{1}{n}}.$$

As each term is strictly increasing, so is λ_n.

Lemma 2 *Let $\lambda \geqslant 1, b > 0$, and $c_1, c_2, \ldots, c_m$ be nonnegative real numbers. Then*

$$\lambda \sum_{i=1}^{m} \sqrt{c_i + b} - \sum_{i=1}^{m} \sqrt{c_i} \geqslant \lambda \sqrt{\sum_{i=1}^{m} c_i + b} - \sqrt{\sum_{i=1}^{m} c_i}.$$

Proof of Lemma 2 Obviously, $\sum_{i=1}^{m} \sqrt{c_i + b} \geqslant \sqrt{\sum_{i=1}^{m} c_i + b}$, and we find

$$\left(\lambda \sum_{i=1}^{m} \sqrt{c_i + b} - \sum_{i=1}^{m} \sqrt{c_i} \right) - \left(\lambda \sqrt{\sum_{i=1}^{m} c_i + b} - \sqrt{\sum_{i=1}^{m} c_i} \right)$$

$$\geqslant \sum_{i=1}^{m} \left(\sqrt{c_i + b} - \sqrt{c_i} \right) - \left(\sqrt{\sum_{i=1}^{m} c_i + b} - \sqrt{\sum_{i=1}^{m} c_i} \right)$$

$$= \sum_{i=1}^{m} \frac{b}{\sqrt{c_i + b} + \sqrt{c_i}} - \frac{b}{\sqrt{\sum_{i=1}^{m} c_i + b} + \sqrt{\sum_{i=1}^{m} c_i}} \geqslant 0.$$

Lemma 3 *Let $\lambda \geqslant \lambda_2, b < a \leqslant c$ be positive real numbers. Then*

$$\lambda(\sqrt{a-b} + \sqrt{c+b}) - \sqrt{a} - \sqrt{c} > \lambda \sqrt{c-a+2b} - \sqrt{b} - \sqrt{c-a+b}.$$

Proof of Lemma 3 The inequality is equivalent to

$$\lambda \sqrt{a-b} + \lambda(\sqrt{c+b} - \sqrt{c-a+2b}) > (\sqrt{a} - \sqrt{b}) + (\sqrt{c} - \sqrt{c-a+b})$$

$$\Leftrightarrow \lambda + \frac{\lambda \sqrt{a-b}}{\sqrt{c+b} + \sqrt{c-a+2b}} > \frac{\sqrt{a-b}}{\sqrt{a} + \sqrt{b}} + \frac{\sqrt{a-b}}{\sqrt{c} + \sqrt{c-a+b}}.$$

On one hand,

$$\lambda > 1 > \frac{\sqrt{a-b}}{\sqrt{a} + \sqrt{b}}. \tag{10}$$

On the other hand, $\lambda \geqslant \lambda_2 = \sqrt{2}$, $\sqrt{c+b} + \sqrt{c-a+2b} < \sqrt{c+c} + \sqrt{2(c-a)+2b} = \sqrt{2} \cdot (\sqrt{c} + \sqrt{c-a+b})$, and thus

$$\frac{\lambda \sqrt{a-b}}{\sqrt{c+b} + \sqrt{c-a+2b}} > \frac{\sqrt{a-b}}{\sqrt{c} + \sqrt{c-a+b}}. \tag{11}$$

By (10) + (11), the lemma is verified.

Return to the original problem. Taking $a_1 = \cdots = a_{n-1} = b = 1, a_n = -(n-1)$, we obtain

$$\lambda\sqrt{n} \geqslant n - 1 + \sqrt{n-1} \Rightarrow \lambda \geqslant \frac{n - 1 + \sqrt{n-1}}{\sqrt{n}} = \lambda_n.$$

It is required to show λ_n is indeed the smallest value, that is, for any real numbers $a_1, a_2, \ldots, a_n$, and b,

$$\lambda_n \sum_{i=1}^{n} \sqrt{|a_i - b|} + \sqrt{n \left| \sum_{i=1}^{n} a_i \right|} \geqslant \sum_{i=1}^{n} \sqrt{|a_i|}. \qquad (12)$$

Let $t = \frac{1}{n}\sum_{i=1}^{n} a_i$ be the average, $a_i' = a_i - t (i = 1, 2, \ldots, n), b' = b - t$. Note that

$$\sqrt{|t|} + \sqrt{|a_i'|} \geqslant \sqrt{|t| + |a_i'|} \geqslant \sqrt{|t + a_i'|} = \sqrt{|a_i|},$$

$$\sqrt{n \left| \sum_{i=1}^{n} a_i \right|} = n\sqrt{|t|} \geqslant \sum_{i=1}^{n} \sqrt{|a_i|} - \sum_{i=1}^{n} \sqrt{|a_i'|}.$$

Also, $|a_i - b| = |a_i' - b'|$. Now, to prove (12), it suffices to prove

$$\lambda_n \sum_{i=1}^{n} \sqrt{|a_i' - b'|} \geqslant \sum_{i=1}^{n} \sqrt{|a_i'|}, \text{ in which } a_1' + \cdots + a_n' = 0. \qquad (13)$$

Assume $b > 0$. By Lemmas 1 and 2 (let $c_i = |a_i'|$ for $a_i' < 0$), we can combine all negative terms in $a_1', a_2', \ldots, a_n'$ into one term. (If the number of terms decreases to $m < n$, use λ_m instead of λ_n; if $b < 0$, combine all positive terms.) To prove (13), we only need: for any nonnegative real numbers $a_1, a_2, \ldots, a_{n-1}, b$, (drop the prime symbols for convenience)

$$\lambda_n \left(\sqrt{c + b} + \sum_{i=1}^{n-1} \sqrt{|a_i - b|} \right) \geqslant \sqrt{c} + \sum_{i=1}^{n-1} \sqrt{a_i}, \text{ in which } c = \sum_{i=1}^{n-1} a_i > 0. \qquad (14)$$

If $b \geqslant \frac{c}{n-1}$, then

$$\lambda_n \left(\sqrt{c + b} + \sum_{i=1}^{n-1} \sqrt{|a_i - b|} \right) \geqslant \lambda_n \sqrt{c + \frac{c}{n-1}} = \sqrt{c} + \sqrt{(n-1)c}$$

$$= \sqrt{c} + \sqrt{(n-1)\sum_{i=1}^{n-1} a_i} \geqslant \sqrt{c} + \sum_{i=1}^{n-1} \sqrt{a_i},$$

and (14) follows.

If $b < \frac{c}{n-1} = \frac{1}{n-1}\sum_{i=1}^{n-1}a_i$, then $b < a_j$ for some j. According to lemma 3, by changing a_j and c to b and $c - a_j + b$ respectively, the difference between the two sides of ⑭ will decrease. After a finite number of adjustments, it will be $b \geqslant \frac{c}{n-1}$, and ⑭ follows.

In conclusion, the minimum value of is $\lambda = \lambda_n = \frac{n-1+\sqrt{n-1}}{\sqrt{n}}$.

Test IV, Second Day
(8 am–12:30 pm; 30 March 2023)

4 Find all functions $f: \mathbf{Z} \to \mathbf{Z}$, such that for any integers a, b, and c,

$$2f(a^2 + b^2 + c^2) - 2f(ab + bc + ca)$$

$$= (f(a - b))^2 + (f(b - c))^2 + (f(c - a))^2. \tag{$*$}$$

(Contributed by Qu Zhenhua)

Solution Either $f(m) = 0, \forall m \in \mathbf{Z}$; or $f(m) = m, \forall\, m \in \mathbf{Z}$.

In $(*)$,

- Taking $a = b = c = 0$: $3(f(0))^2 = 0$, and hence $f(0) = 0$.
- Taking $a = 1, b = c = 0$: $2f(1) = (f(1))^2 + (f(-1))^2$, or

$$(f(1) - 1)^2 + (f(-1))^2 = 1.$$

It must be either $f(1) = 1, f(-1) = \pm 1$ or $f(-1) = 0, f(1) = 0, 2$.

- Taking $a = b = 1, c = 0$: $2f(2) - 2f(1) = (f(1))^2 + (f(-1))^2 = 2f(1)$, yielding $f(2) = 2f(1)$.
- Taking $a = 1, b = 0, c = -1$: $2f(2) - 2f(-1) = (f(2))^2 + 2(f(-1))^2$, or $(f(2) - 1)^2 + 2(f(-1) + \frac{1}{2})^2 = 1.5$, and thus

$$f(2) = f(-1) = 0, \quad \text{or} \quad f(2) = 2, \quad f(-1) = -1,$$

which further implies that

$$f(1) = f(-1) = 0, \quad \text{or} \quad f(1) = 1, \quad f(-1) = -1.$$

Let $g(k) = (f(k))^2 - (f(-k))^2$. We have $g(-k) = -g(k)$ and

$$g(0) = g(1) = 0.$$

In $(*)$, if (a, b, c) is replaced by (b, c, a), then the left side does not change while the right side changes by $g(a-b)+g(b-c)+g(c-a)$. Taking $(a, b, c) = (k, 1, 0)$, $g(k - 1) + g(1) + g(-k) = 0$, or $g(k) = g(k - 1) + g(1)$. Hence, $g(k) = 0$ for all k, which means $f(-k) = \pm f(k)$.

Taking $(a, b, c) = (k, -1, 0)$ and $(a, b, c) = (k, 1, 0)$, the difference of which gives

$$2(f(k) - f(-k)) = (f(k+1))^2 - (f(k-1))^2. \qquad \text{⑮}$$

In the first case, $f(1) = f(-1) = 0$.

Claim: For every nonnegative integer k, $f(k) = f(-k) = 0$. By induction, $k = 0, 1$ have been verified. Suppose $f(\pm k) = 0, f(\pm(k-1)) = 0$. From ⑮, it follows $(f(k+1))^2 = 0$, or $f(k+1) = 0$, and $f(-k-1) = \pm f(k+1) = 0$, completing the inductive proof.

Therefore, in the first case, $f(m) = 0, \forall\, m \in \mathbf{Z}$.

In the second case, $f(1) = 1, f(-1) = -1, f(2) = 2f(1) = 2$.

In (∗), taking $(a, b, c) = (k, 1, -1)$ and $(a, b, c) = (k, 1, 1)$, the difference of which gives

$$2f(2k+1) - 2f(-1) = (f(-1-k))^2 - (f(1-k))^2 + (f(2))^2$$

$$\Rightarrow f(2k+1) = \frac{1}{2}(f(k+1))^2 - \frac{1}{2}(f(k-1))^2 + 1. \qquad \text{⑯}$$

Taking $(a, b, c) = (k, 2, 0)$ and $(a, b, c) = (k, -2, 0)$, the difference of which gives

$$2f(2k) - 2f(-2k) = (f(k+2))^2 - (f(k-2))^2. \qquad \text{⑰}$$

In ⑯, taking $k = 1$, we get $f(3) = 3$.

In ⑰, taking $k = 2$, $(f(4))^2 = 2(f(4) - f(-4))$. If $f(4) = 0$, then $k = 3$ in ⑮ gives $f(3) - f(-3) = -2$, implying $f(-3) = 5 \neq \pm f(3)$, a contradiction. Hence, $f(4) \neq 0$. As $f(-4) = \pm f(4)$, it must be $f(-4) = -f(4)$, so $(f(4))^2 = 4f(4), f(4) = 4$.

Claim: For every nonnegative integer m, $f(m) = m$. By induction, $m = 0, 1, 2, 3, 4$ have been verified. Suppose $f(k) = k$ for $k = 0, 1, 2, \ldots, m-1$.

If $m = 2k + 1 \geqslant 5$, by ②,

$$f(2k+1) = \frac{1}{2}(f(k+1))^2 - \frac{1}{2}(f(k-1))^2 + 1 = 2k + 1.$$

If $m = 2k \geqslant 6$, by ③,

$$f(2k) - f(-2k) = \frac{1}{2}(f(k+2))^2 - \frac{1}{2}(f(k-2))^2 = 4k,$$

and $f(-2k) = \pm f(2k)$. It must be $f(-2k) = -f(2k)$, so $2f(2k) = 4k, f(2k) = 2k$. This completes the induction.

Finally, for positive integer m, taking $k = m$ in ⑮ yields

$$2(f(m) - f(-m)) = (f(m+1))^2 - (f(m-1))^2 = 4m \Rightarrow f(-m) = -m.$$

Therefore, in the second case, $f(m) = m, \forall\, m \in \mathbf{Z}$.

It is straightforward to check that $f(m) = 0$ and $f(m) = m$ are solutions of $(*)$.

⑤ Given a prime number p and a real number $\lambda \in (0,1)$. Let s, t be positive integers such that $s \leqslant t < \frac{\lambda}{12}p$. Let S, T be sets of s and t consecutive integers respectively, let k be an integer, such that

$$|(x, y) \in S \times T : kx \equiv y \pmod{p}| \geqslant 1 + \lambda s,$$

Show that there exist integers a and b, satisfying

$$ak \equiv b \pmod{p}, \quad 0 < a \leqslant \frac{1}{\lambda}, \quad 0 \leqslant |b| \leqslant \frac{t}{\lambda s}.$$

(Contributed by Ai Yinghua)

Solution Let the solutions of the congruence equation $kx \equiv y \pmod{p}$ be $(x_1, y_1), \ldots, (x_n, y_n)$ in $S \times T$, where $n \geqslant 1 + \lambda s \geqslant 1$. So, S, T are nonempty, $n \geqslant 2$. As T is contained in a complete system of residues mod p, we see that x_i's are all distinct. Let $x_1 < x_2 < \cdots < x_n$, $u_i = x_i - x_1, v_i = y_i - y_1$. Then

$$ku_i \equiv v_i \pmod{p}, \quad 0 \leqslant u_i \leqslant s, \quad |v_i| \leqslant t.$$

Define $a = \frac{u_2}{\gcd(u_2, v_2)}, b = \frac{v_2}{\gcd(u_2, v_2)}$. We have $\gcd(a, b) = 1$ and

$$0 < a \leqslant s, \quad |b| \leqslant t, \quad ka \equiv b \pmod{p}.$$

Since $ku_i \equiv v_i \pmod{p}$, it follows

$$bu_i \equiv kau_i \equiv av_i \pmod{p}.$$

Let $bu_i = av_i + pw_i$. We have

$$|w_i| \leqslant \frac{|b|s + at}{p}.$$

Let $L = \frac{|b|s + at}{p}$.

(1) If $L < 1$, then all w_i's are zero, and thus $bu_i = av_i$. From $\gcd(a, b) = 1$, we see that every u_i is a multiple of a. As $u_i \in [0, s]$ are all distinct, it follows that

$$s \geqslant \max u_i \geqslant a(n-1) \geqslant a\lambda s,$$

yielding $a \leqslant \frac{1}{\lambda}$. Now, from

$$at \geqslant \max |av_i| = \max |bu_i| \geqslant |b| \cdot (a\lambda s),$$

we obtain $|b| \leqslant \frac{t}{\lambda s}$.

(2) If $L \geqslant 1$, then w_i can take at most $2L+1 \leqslant 3L$ integer values in $[-L, L]$. By the pigeonhole principle, at least $\frac{n}{3L}$ of w_i's take the same value. Let E be the set of the subscripts of these identical w_i's and $u_{i_0} = \min\{u_i : i \in E\}$. Since

$$b(u_i - u_{i_0}) = a(v_i - v_{i_0}) + p(w_i - w_{i_0}) = a(v_i - v_{i_0}), \quad \forall\, i \in E,$$

we have $a|(u_i - u_{i_0})$; also, $0 \leqslant u_i - u_{i_0} = x_i - x_{i_0} \leqslant s$, and hence

$$s \geqslant \max_{i \in E}(u_i - u_{i_0}) \geqslant a(|E| - 1), \tag{18}$$

combined with $|v_i - v_{i_0}| = |y_i - y_{i_0}| \leqslant t$, it follows that

$$at \geqslant \max(a|v_i - v_{i_0}|) = |b|\max_{i \in E}(u_i - u_{i_0}) \geqslant |b|(a(|E| - 1)). \tag{19}$$

Note that

$$L = \frac{|b|s + at}{p} \leqslant \frac{2st}{p} < \frac{\lambda s}{6}.$$

We get $1 < \frac{\lambda s}{6L}$, and furthermore

$$|E| - 1 \geqslant \frac{\lambda s}{3L} - \frac{\lambda s}{6L} = \frac{\lambda s}{6L}.$$

Applying it to (18) and (19), we find

$$\lambda a \leqslant 6L, \quad \lambda|b|s \leqslant 6Lt,$$

which further imply that

$$\lambda p = \frac{\lambda|b|s + \lambda at}{L} \leqslant 12t,$$

but this contradicts the condition $t < \frac{\lambda p}{12}$.

Consequently, $L \geqslant 1$ cannot occur, and it must be $L < 1$. By the argument in ①, the conclusion follows.

⑥ Given positive integer n, find the smallest positive integer k such that: for any $2n \times 2n$ grid with k black unit squares (the other squares are white), one can always make all squares black by a finite number of operations as follows:

(operation A) take a 2×2 square with exactly 1 white square and color it black;

(operation B) take a 2×2 square with exactly 2 white squares and change the colors of all four squares (black to white, white to black).
(Contributed by Xiao Liang and Fu Yunhao)

Solution The minimum of k is $n^2 + n + 1$.

First, we give a grid configuration with $n^2 + n$ black squares such that operation A can never be performed, and therefore it cannot be all black after a finite number of operations. This implies $k \geqslant n^2 + n + 1$.

Color all anti-diagonal squares black; color (i, j), i, j both odd and $i+j < 2n$ (above the anti-diagonal) black; color $(i, j), i, j$ both even and $i + j > 2n + 2$ (below the anti-diagonal) black. Figure 7.12 shows the configuration for $n = 4$, where marked squares are colored black.

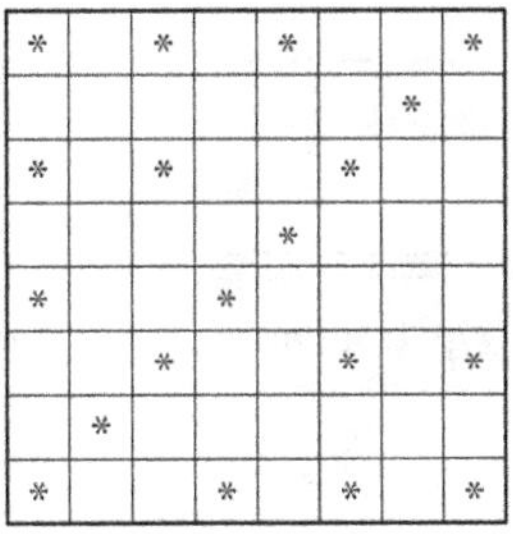

Fig. 7.12

Clearly, the grid consists of $n^2 + n$ black squares, and it satisfies

Condition $(*)$: for any black squares (i, j_1) and (i, j_2) in the same row, $j_1 < j_2$, there exists $j_1 < j < j_2$ such that $(i - 1, j), (i, j), (i + 1, j)$ are all white; for any black squares (i_1, j) and (i_2, j) in the same column, $i_1 < i_2$, there exists $i_1 < i < i_2$ such that $(i, j-1), (i, j), (i, j+1)$ are all white. Here, all squares outside the grid are assumed to be white.

Note that if a configuration satisfies ($*$), then any 2×3 (or 3×2) rectangle contains at most 2 black squares, as otherwise either (i) two black squares have a common side; or (ii) a white square shares common sides with 3 black squares, both violating ($*$).

Next, we show that only operation B can be performed for such a configuration, and condition ($*$) is satisfied for all subsequent configurations, in particular, no two black squares can have a common side (called '*adjacent*'). Once verified, the grid cannot be all black as the number of black squares never increases. Clearly, the initial configuration satisfies ($*$). Suppose the assertion is true after some operation B (since A cannot be performed). We only need to show the new configuration satisfies ($*$). By symmetry, let (i, j_1) and (i, j_2) be two black squares. There are two cases.

- **Case 1:** (i, j_1) and (i, j_2) are black before the operation. Then, there exists $j_1 < j < j_2$, such that $(i-1, j), (i, j), (i+1, j)$ are all white, as shown in Fig. 7.13. Note that the rectangle $[i-1, i+1] \times [j_1+1, j-1]$ contains at most $j - j_1 - 1$ black squares (if $j - j_1 - 1$ is even, divide it into 3×2 rectangles, each containing at most 2 black squares; if $j - j_1 - 1$ is odd, add the squares $(i-1, j_1), (i, j_1), (i+1, j_1)$ then divide). Likewise, the rectangle $[i-1, i+1] \times [j+1, j_2-1]$ contains at most $j_2 - j - 1$ black squares. Hence, $T = [i-1, i+1] \times [j_1+1, j_2-1]$ contains at most $j_2 - j_1 - 2$ black squares.

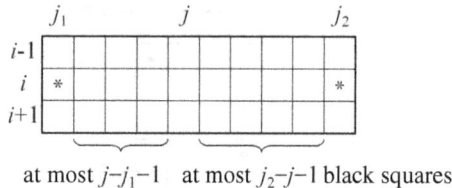

at most $j - j_1 - 1$ at most $j_2 - j - 1$ black squares

Fig. 7.13

After the operation, if the number of black squares in T does not increase, then T still has a column of white squares, satisfying ($*$); if the number of black squares in T increases, as no black squares can be adjacent before the operation, the new one must be a corner square of T. Then, one of the four squares

$$(i-1, j_1), \quad (i+1, j_1), \quad (i-1, j_2), \quad (i+1, j_2)$$

is black before the operation, meaning that two black squares are adjacent, but this is a contradiction.

- **Case 2:** One of (i, j_1) and (i, j_2) is a new black square (not both, because the two new black squares are in different rows and columns). By symmetry, suppose (i, j_2) is the new black square and the latest operation B is performed on a 2×2 square in rows $i-1$ and i. If it changes $(i-1, j_2)$ and (i, j_2-1) from black to white, then apply $(*)$ to (i, j_1) and (i, j_2-1) before the operation to find $j_1 < j < j_2 - 1$ such that $(i, j-1), (i, j), (i, j+1)$ are all white, and they remain white after the operation; if it changes $(i-1, j_2)$ and (i, j_2+1) from black to white, then apply $(*)$ to (i, j_1) and (i, j_2+1) before the operation to find $j_1 < j < j_2 + 1$ such that $(i, j-1), (i, j), (i, j+1)$ are all white, and $j \neq j_2$, j still satisfies $(*)$ after the operation.

Now we prove $k = n^2 + n + 1$ satisfies the problem condition. It suffices to Prove that if the grid consists of at least $n^2 + n + 1$ black squares and at least 1 white square, then after a sequence of operations B, an operation A can be performed. Suppose this is not true. We need a lemma first.

Lemma *Let x_i be the number of black squares in row i. If $x_{i-1} + 2x_i + x_{i+1} \geqslant 2n + 2$, then by a sequence of operations B in the rectangle $T = [i-1, i+1] \times [1, 2n]$, we can make two adjacent black squares in a row. Here, $x_0 = x_{2n+1} = 0$.*

Proof of Lemma First, we assert that we can always perform a sequence of operations B in T to make two adjacent black squares while maintaining the values x_{i-1}, x_i, x_{i+1} in the process. Suppose this is not true. Then no two black squares have a common side in the current configuration. Let the black squares in row i occupy columns $j_1, \ldots, j_r$ ($j_1 < \cdots < j_r$). If for $k = 1, \ldots, r-1$, columns $j_k+1, \ldots, j_{k+1}-1$ of T contain at most $j_{k+1}-j_k-2$ black squares, the first $j_1 - 1$ columns of T contain at most $j_1 - 1$ black squares, and the last $2n - j_r$ columns of T contain at most $2n - j_r$ black squares, then

$$x_{i-1} + 2x_i + x_{i+1} \leqslant 2r + (j_1 - 1) + (2n - j_r) + \sum_{k=1}^{r-1} (j_{k+1} - j_k - 2) = 2n + 1$$

leads to a contradiction.

If the first $j_1 - 1$ columns contain more than $j_1 - 1$ black squares or the last $2n - j_r$ columns contain more than $2n - j_r$ black squares, then some 3×2 rectangle holds at least 3 black squares: clearly, no two are adjacent. Perform an operation B to get two adjacent black squares, and x_{i-1}, x_i, x_{i+1} remain the same.

If for some $k \in \{1, \ldots, r-1\}$, columns $j_k + 1, \ldots, j_{k+1} - 1$ of T contain $j_{k+1} - j_k - 1$ or more black squares, then two possible situations can occur: (i) two adjacent columns (3×2 rectangle) hold 3 black squares, then perform an operation B to get adjacent black squares; (ii) otherwise, each of columns $j_k + 1$ and $j_{k+1} - 1$ contains at most one black square, by two operations B on columns $j_k, j_k + 1$ and columns $j_{k+1} - 1, j_{k+1}$, we can make $(i, j_k + 1)$, $(i, j_{k+1} - 1)$ black, with $j_{k+1} - j_k - 3$ or more black squares in between. Continue until we reach (i) or two adjacent black squares.

Now we have obtained two adjacent black squares. If they are in the same row, the lemma conclusion follows. If they are in the same column, perform operations B to move them horizontally until one becomes adjacent to another black square in the same row. If there are no other black squares in these rows, then the third row of T contains at least $(2n+2) - 2 \cdot 1 - 1 = 2n - 1$ black squares, two of which must be adjacent. Either way, we have achieved two horizontally adjacent black squares, and the lemma is proved.

Return to the original problem. Note that

$$x_1 + \sum_{i=1}^{n} (x_{2i-1} + 2x_{2i} + x_{2i+1}) \geqslant 2(n^2 + n + 1),$$

where $x_{2n+1} = 0$. If $x_1 \geqslant n+1$, there are two adjacent black squares in the first row; otherwise $x_1 \leqslant n$, then for some $i \in \{1, 2, \ldots, n\}$,

$$x_{2i-1} + 2x_{2i} + x_{2i+1} \geqslant \left\lfloor \frac{2n^2 + n + 2}{n} \right\rfloor \geqslant 2n + 2,$$

by the lemma we can perform operations B in rows $2i - 1, 2i, 2i + 1$ to get adjacent black squares in the same row.

Suppose $(2i, j_1), (2i, j_1 + 1)$ are black. Let u, v be maximal such that every column from $j_1 - u$ to $j_1 + v$ contains a black square. If a column contains two or more black squares, say column $j_1 - s$. By a sequence of operations B, we can move the black pair $(2i, j_1), (2i, j_1 + 1)$ vertically until the one in column j_1 is adjacent to the black square in column $j_1 - 1$, and then move this pair vertically until the one in column $j_1 - 1$ is adjacent to the black square in column $j_1 - 2$, and so on, until we get a black pair in columns $j_1 - s, j_1 - s + 1$. Eventually, we reach a 2×2 square, with at least 3 black squares (this could happen earlier, though). If an operation A can be performed, then the conclusion (before the lemma) follows. Otherwise, all four squares are black, and we move them horizontally or vertically until they touch other black squares. Suppose at this moment an operation A still cannot be performed, then it must be a 2×3 or 3×2 black rectangle. Consider the 3 consecutive squares say a, b, c adjacent to the black rectangle: if all are black, turn to the next row or column (if this is always the case, the grid is all black!); if one or two of them are black, we can perform an operation A right away; if none is black, by changing a, b to black, we can get a 2×2 square with exactly 3 black squares and perform an operation A.

Henceforth, we assume that every column from $j_1 - u$ to $j_1 + v$ contains exactly one black square. Let y_j be the number of black squares in column j. Then, either there are two adjacent black squares in the first column, or for some $1 \leqslant j \leqslant n$, $y_{2j-1} + 2y_{2j} + y_{2j+1} \geqslant 2n + 2$. Since $y_{j_1-u} = y_{j_1-u+1} = \cdots = y_{j_1+v} = 1$, $y_{j_1-u-1} = y_{j_1+v+1} = 0$ it follows that

$$2j \notin \{j_1 - u - 1, j_1 - u, \ldots, j_1 + v + 1\}.$$

We adopt a similar method as in the lemma to columns $2j - 1, 2j, 2j + 1$ to obtain two vertically adjacent black squares, in the meantime the operations do not change the two horizontally adjacent black squares that we obtained earlier. Now, move the vertical pair horizontally and the horizontal pair vertically so that they 'meet', and we reach a 2×2 square with at least 3 black squares. The case of four black squares can be handled by the previous argument. The proof is now complete.

Therefore, the smallest $k = n^2 + n + 1$.

Remark 1 For the same grid with $n^2 + n$ black squares as described, here are another two proofs that operations A can never be performed.

1. Edited from Liang Xingjian's solution. Divide the grid into n^2 2×2 blocks. We show that (a) black squares in a block are never adjacent (with a common side); (b) black squares in adjacent blocks (with a common side) always have no common vertex.

   The initial configuration satisfies conditions (a) and (b). Suppose the same is true for the current configuration. If we perform an operation within a block, by (a) it must be operation B, and by (b) the 8 surrounding squares must be all white (see Fig. 7.14, where the block consists of the four central squares). It is not difficult to see that conditions (a) and (b) still hold after the operation.

   If the operation is not performed within a block, then it is not in two adjacent blocks, either, by (b). So the operation is on four squares all in different blocks; it must be operation B. Again, by (a) and (b), in the figure (now there are four blocks), the squares without question

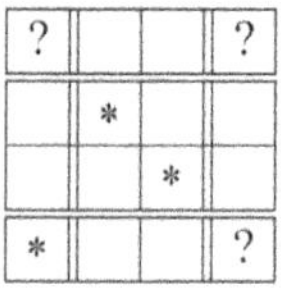

Fig. 7.14

marks are all white, and thus conditions (a) and (b) still hold after the operation.

2. Edited from Sun Qi'ao and Liu Jinchang's solutions. Color the grid squares alternately red and blue. Two black squares are adjacent if and only if they share a common vertex. The key is to show the current configuration always satisfies (before and after operation B)

   (a) No two black squares have a common side (so each connected component is either all red or all blue);

   (b) Every red connected component (can be one square) is a chain from upper left to lower right;

   (c) Every blue connected component (can be one square) is a chain from upper right to lower left.

   The proof involves similar case-by-case discussions and is left to interested readers.

Remark 2   The following elegant proof that $k = n^2 + n + 1$ satisfies the problem condition is provided by Jiang Cheng.

Lemma   *If there are more than $\frac{(m+1)(n+1)}{4}$ black squares in an $m \times n$ grid, then a sequence of operations can make two black squares adjacent (with a common side).*

Proof of Lemma   Suppose it is not always possible. Among all such configurations, choose the one with the smallest

$$\sum_{(x,y) \text{ is black}} (x+y)^3 \tag{$*$}$$

value (the summation is over all black squares in this configuration). Then, there are no two black squares like (x, y) and $(x + 1, y + 1)$, as we can perform an operation B to get black squares $(x, y+1)$ and $(x+1, y)$, which give a smaller sum ($*$). Likewise, there are no black squares

$$(x, y + 1), \quad (x + 1, y), \quad (x + 2, y + 2),$$

as we can perform two operations B to get black squares $(x, y + 1)$, $(x + 1, y), (x + 2, y + 2)$ to make the sum ($*$) smaller.

Now consider the expanded grid $(i,j)(1 \leqslant i \leqslant m+1; 1 \leqslant j \leqslant n+1)$, whose last row and last column are assumed all white. Suppose, along the line $x+y = a+b+t$, there are $t+1$ consecutive black squares $(a+t,b),\ldots,(a,b+t)$. Since this configuration has the smallest sum $(*)$, by the above argument, none of the following $3t+3$ squares

$$(a+t+1,b),\ldots, \quad (a,b+t+1),$$

$$(a+t+1,b+1),\ldots, \quad (a+1,b+t+1),$$

$$(a+t+1,b+2),\ldots, \quad (a+2,b+t+1),$$

can be black (some of them can be in the last row or column; if $t = 0$, the squares to the right, bottom, bottom right cannot be black). We see that for black squares on different lines $x+y = c_i$, or different black sequences along the same line, they all correspond to distinct non-black squares. Hence, in the expanded grid, each black square correspond to three white square, and the conclusion follows.

Return to the original problem. If we can always make a 2×2 block with three black squares, then eventually the whole grid is filled with black squares. According to the lemma, there is a pair of adjacent black squares, say in the same row. By performing operations B, they move freely in the two columns. Then each neighboring column must contain exactly one black square, until on the two sides we find two all-white columns (by an argument similar to the official proof). This implies that the whole grid can be separated into a few sub-grids, either the total number of black squares is less than or equal to $n^2 + n$, or there is a pair of vertically adjacent black squares, which can be moved to meet the horizontal black pair to generate new black squares.

8

The 14th Romanian Master of Mathematics Competition

2023 (Bucharest, Romania)

Organized by the Romanian Mathematical Society and hosted by the "Tudor Vianu" national college of computer science, the 14th Romanian Master of Mathematics Competition (RMM) was held in Bucharest from 27th February to 4th March 2023. The RMM provides a fantastic opportunity for young students from different countries and regions to make new friends, demonstrate their abilities in mathematics, exchange knowledge and to enhance cross-cultural contacts, and learn about Romanian culture. Due to the impact of the pandemic, the Chinese team did not participate in this competition.

The 14th Romanian Master of Mathematics Competition featured 15 teams from 12 countries, with a total of 90 students participating. After two days of exams, eight students won gold medals, 13 students won silver medals, and 26 students won bronze medals. The teams from the USA, Israel, Romania, Serbia, Bulgaria and Hungary achieved the top six positions in the overall team rankings.

First Day
1st March 2023

① Determine all prime numbers p and all positive integers x and y satisfying

$$x^3 + y^3 = p(xy + p).$$

Solution 1 For prime numbers p and positive integers x and y that satisfy the given condition, since

$$(x + y)(x^2 - xy + y^2) = p(xy + p),$$

we have $p \mid (x + y)(x^2 - xy + y^2)$.

(1) If $p \mid x + y$, then $\frac{x+y}{p} = \frac{xy+p}{x^2-xy+y^2}$.

If $p = x+y$, then $x^2 - xy + y^2 = xy + p$, so $p = (x-y)^2$, which contradicts p being a prime number.

If $p < x + y$, then $\frac{x+y}{p} \geqslant 2$, so

$$\frac{xy + p}{x^2 - xy + y^2} \geqslant 2, p - xy \geqslant 2(x - y)^2 \geqslant 0.$$

Therefore, $x + y \geqslant 2p \geqslant 2xy$, which contradicts x and y being positive integers.

(2) If $p \nmid x + y$, then $p \mid x^2 - xy + y^2$, so

$$\frac{x^2 - xy + y^2}{p} = \frac{xy + p}{x + y}$$

is a positive integer. Thus,

$$\frac{3(x^2 - xy + y^2 + 3(xy + p))}{p + 3(x + y)} = \frac{3(x + y)^2 + 9p}{p + 3(x + y)} = x + y + \frac{(9 - x - y)p}{p + 3(x + y)}$$

is a positive integer.

Therefore, $p + 3(x + y) \mid (9 - x - y)p$, so $p \mid p + 3(x + y)$ or

$$p + 3(x + y) \mid 9 - x - y.$$

If $p \mid p + 3(x + y)$, since $p \nmid x + y$, then $p = 3$, so $x + y = 2$ or

$$x + y \geqslant 4.$$

If $x + y = 2$, then $x = y = 1$, which does not satisfy the condition.

If $x + y \geqslant 4$, then $4(x - y)^2 \leqslant 9 - xy$, so $x = y = 2$ or $x = y = 3$, both of which do not satisfy the condition.

If $p + 3(x + y) \mid 9 - x - y$, then since $p + 3(x + y) > |9 - x - y|$, there is

$$9 - x - y = 0, x + y = 9.$$

Substituting the eight possible values of $x + y = 9$, it can be easily seen that all the triples (x, y, p) that satisfy the conditions are $(1, 8, 19), (8, 1, 19), (2, 7, 13), (7, 2, 13), (4, 5, 7), (5, 4, 7)$.

Solution 2   Let $s = x + y$. The given equation in the question transforms to $s(s^2 - 3xy) = p(p + xy)$, that is,

$$xy = \frac{s^3 - p^2}{3s + p},$$

$$s^2 \geqslant 4xy = \frac{4(s^3 - p^2)}{3s + p}.$$

Therefore,

$$(s - 2p)(s^2 + sp + 2p^2) \leqslant 4(p^2 - p^3) < 0,$$

and hence $s < 2p$.

(1) If $p \mid s$, then $s = p$, so $xy = \frac{p(p-1)}{4}$. Since $t^2 - pt + \frac{p(p-1)}{4} = 0$ has no solutions, the integers x and y that satisfy the condition do not exist.

(2) If $p \nmid s$, transform the given equation in the question to

$$27xy = (9s^2 - 3sp + p^2) - \frac{p^2(p + 27)}{3s + p}.$$

Since both sides of the equation are integers and $p \nmid s$, we have $3s + p \mid p + 27, 3s + p \mid 27 - 3s$.

If $s \neq 9$, then $|3s - 27| \geqslant 3s + p$, so $27 - 3s \geqslant 3s + p, s \leqslant 4$. After verification one by one, it is known that there are no solutions.

If $s = 9$, it can be verified that all the triples (x, y, p) that satisfy the conditions are $(1, 8, 19), (8, 1, 19), (2, 7, 13), (7, 2, 13), (4, 5, 7), (5, 4, 7)$.

2   Fix an integer $n \geqslant 3$. Let $\mathcal{S}$ be a set of n points in the plane, no three of which are collinear. Given different points A, B, C in $\mathcal{S}$, $\triangle ABC$ is nice for AB if $S_{\triangle ABC} \leqslant S_{\triangle ABX}$ for all X in $\mathcal{S}$ different from A and B. (Note that for a segment AB there could be several nice triangles.) A triangle is *beautiful* if its vertices are all in $\mathcal{S}$ and it is *nice for* at least two of its sides.

Prove that there are at least $\frac{1}{2}(n - 1)$ beautiful triangles.

Solution 1   Construct a graph with the points in $\mathcal{S}$ as vertices: For each beautiful triangle, it is nice for at least two of its three sides. By selecting two of the edges to connect in the graph, then these connected edges make the three vertices of this beautiful triangle connected. Denote the graph formed by connecting lines in this way as G.

If the graph G is not connected, then we can divide the vertices of graph G into two parts, X and Y, with no edges connecting them. Consider the set K of triangles with all vertices containing points in both X and Y.

Denote the triangle with the smallest area as $\triangle ABC$. We might as well assume $A, B \in X, C \in Y$, and then the triangles containing the side AC or side BC are all in the set K. Since $\triangle ABC$ has the smallest area, $\triangle ABC$ is *nice for AC* and *BC*. Therefore, $\triangle ABC$ is beautiful. By the rules for connecting vertices in G, it is known that vertices A, B, C are connected, implying there are edges between X and Y, which contradicts the assumption. Hence, graph G is connected.

Therefore, graph G has at least $n-1$ edges. Since each beautiful triangle connects two sides, there are at least $\frac{1}{2}(n-1)$ beautiful triangles.

Solution 2 Construct a graph with the points in S as vertices, denoted as G.

Lemma *If the vertices of graph G are divided into two parts, X and Y, then the triangles containing the side AC or side BC are all in the set K. Denote the triangle with the smallest areas as $\triangle ABC$, and then $\triangle ABC$ is beautiful.*

Proof of the Lemma. We might as well assume $A, B \in X, C \in Y$, and then the triangles containing the side AC or side BC are all in the set K. Since $\triangle ABC$ has the smallest area, $\triangle ABC$ is *nice for AC* and *BC*. Therefore, $\triangle ABC$ is beautiful.

Now we divide the vertices of graph G into two parts, X and Y, where X contains only one point and Y contains $n-1$ points. By the lemma, there exists a triangle whose vertices contain points in X and Y that is beautiful. Put the vertices of this triangle that are in Y into X. This adds two vertices to X and subtracts two vertices from Y. The vertex partition of the changed graph G is still referred to as X and Y. Repeat the above operation. The beautiful triangles that appear in each operation are mutually different. At the same time, at most two vertices are added to X after each operation, and at most two vertices are reduced in Y. The operation is continued until there are no vertices in Y. This operation is performed at least $\frac{n-1}{2}$ times, which means that at least $\frac{n-1}{2}$ different beautiful triangles have appeared in total. The proof is completed.

③ Let $n \geqslant 2$ be an integer, and let f be a $4n$-variable polynomial with real coefficients. Assume that, for any $2n$ points $(x_1, y_1), \ldots, (x_{2n}, y_{2n})$ in the plane,

$$f(x_1, y_1, \ldots, x_{2n}, y_{2n}) = 0$$

if and only if the points form the vertices of a regular $2n$-gon in some order, or are all equal.

Determine the smallest possible degree of f. (*Note*: The degree of polynomial $g(x,y) = 4x^3y^4 + xy + x - 2$ is 7, because $7 = 3 + 4$.)

Solution We will prove in two steps that the smallest possible degree of f is $2n$.

(1) First, we prove that $\deg f \geqslant 2n$.

Let $z_k = x_k + iy_k, k = 1, 2, \ldots, 2n$, and define function g satisfying

$$g(z_1, \overline{z_1}, \ldots, z_{2n}, \overline{z_{2n}})$$
$$= f\left(\frac{z_1 + \overline{z_1}}{2}, \frac{z_1 - \overline{z_1}}{2i}, \ldots, \frac{z_{2n} + \overline{z_{2n}}}{2}, \frac{z_{2n} - \overline{z_{2n}}}{2i}\right).$$

According to the given conditions, $g(z_1, \overline{z_1}, \ldots, z_{2n}, \overline{z_{2n}}) = 0$ if and only if $z_1, z_2, \ldots, z_{2n}$ correspond to the vertices of a regular $2n$-gon or they all coincide. By translating, rotating, and scaling the coordinate system, we can assume that this regular $2n$-gon is centered at the origin and one vertex is $(1, 0)$.

Consider the function

$$G(x) = g(x, \overline{x}, x\omega^2, \overline{x\omega^2}, \ldots, x\omega^{2n-2}, \overline{x\omega^{2n-2}}, \omega, \hat{\omega}, \omega^3, \overline{\omega^3}, \ldots,$$
$$\omega^{2n-1}, \overline{\omega^{2n-1}}),$$

where $\omega = \cos\frac{\pi}{2n} + i\sin\frac{\pi}{2n}$.

Then $G(x) = 0$ if and only if $x = 1, \omega^2, \omega^4, \ldots, \omega^{2n-2}$, that is, $x^n - 1 \mid G(x)$.

Assume there exist z_1, z_2 such that $G(z_1) > 0, G(z_2) < 0$. Consider a path from z_1 to z_2 that does not pass through the zero point. By the intermediate value theorem, there must exist a point z_0 on this path such that $G(z_0) = 0$, which contradicts the assumption that it does not pass through the zero point. Therefore, $G(x)$ is either always non-negative or always non-positive in the complex plane. We might as well assume that $G(x)$ is always non-negative on the complex plane, i.e., $G(x) \geqslant 0$, $\forall x \in C$.

Since the regular $2n$-gon moved to the unit element is chosen arbitrarily, for any $z_1, z_2, \ldots, z_{2n}$, $g(z_1, \overline{z_1}, \ldots, z_{2n}, \overline{z_{2n}}) \geqslant 0$. Hence, every root of $G(x) = 0$ is at least a double root. Therefore, by $x^n - 1 \mid G(x)$, we have $(x^n - 1)^2 \mid G(x)$. So $\deg G \geqslant 2n$, and hence $\deg f \geqslant 2n$.

(2) Next, we prove that there exists an f satisfying the conditions of the problem such that $\deg f = 2n$.

Lemma *Assume* $z_1, z_2, \ldots, z_{2n}$ *are* $2n$ *non-zero complex numbers with the same magnitude, and*

$$z_1^k + z_2^k + \cdots + z_{2n}^k = 0, k = 1, 2, \ldots, n.$$

Then $z_1, z_2, \ldots, z_{2n}$ *are the vertices of a regular* $2n$*-gon centered at the origin.*

Proof of the Lemma Since $z_1, z_2, \ldots, z_{2n}$ are of the same magnitude, we might as well assume that these $2n$ numbers all lie on the unit circle. By Newton's formula, for $k = 1, 2, \ldots, n$, the cyclic sum of any kth power product of $z_1, z_2, \ldots, z_{2n}$ is all equal to 0. Taking the conjugate of $z_1^k + z_2^k + \cdots + z_{2n}^k = 0$, we can get $z_1^{-k} + z_2^{-k} + \cdots + z_{2n}^{-k} = 0, k = 1, 2, \ldots, n$. Therefore, according to Newton's formula, for $k = 1, 2, \ldots, n$, the cyclic sum of any kth power product of $z_1^{-1}, z_2^{-1}, \ldots, z_{2n}^{-1}$ is all equal to 0. By multiplying $z_1 z_2 \cdots z_{2n}$, we know that for $k = 2n - 1, 2n - 2, \ldots, n$, the cyclic sum of any kth power product of $z_1, z_2, \ldots, z_{2n}$ is all equal to 0. Combining with the previous conclusion, we can conclude that for $k = 1, 2, \ldots, 2n - 1$, the cyclic sum of any kth power product of $z_1, z_2, \ldots, z_{2n}$ is all equal to 0. Consider function

$$h(x) = (x - z_1)(x - z_2) \cdots (x - z_{2n}),$$

and it is easy to find that the kth coefficient of $h(x)$ is the cyclic sum of the kth power product of $z_1, z_2, \ldots, z_{2n}$, which equals 0. That is, $h(x) = x^{2n} - C$, where $|C| = 1$. Therefore, $z_1, z_2, \ldots, z_{2n}$ are the vertices of a regular $2n$-gon centered at the origin. The proof of the lemma is complete.

Returning to the original problem, take

$$f(x_1, y_1, \ldots, x_{2n}, y_{2n}) = \sum_{k=1}^{2n} (|\omega_k|^2 - |\omega_1|^2)^2 + \sum_{k=1}^{n} |\omega_1^k + \omega_2^k + \cdots + \omega_{2n}^k|^2,$$

where $\omega_k = z_k - \frac{1}{2n} \sum_{i=1}^{2n} z_i, z_k = x_k + i y_k, k = 1, 2, \ldots, 2n$.

Obviously, if $(x_1, y_1), \ldots, (x_{2n}, y_{2n})$ form the vertices of a regular $2n$-gon in a certain order or they are all the same, $f(x_1, y_1, \ldots, x_{2n}, y_{2n}) = 0$. And when $f(x_1, y_1, \ldots, x_{2n}, y_{2n}) = 0$, there must be

$$|\omega_k|^2 - |\omega_1|^2 = 0, \quad k = 1, 2, \ldots, 2n$$

and

$$\omega_1^k + \omega_2^k + \cdots + \omega_{2n}^k = 0, \quad k = 1, 2, \ldots, n.$$

The former formula ensures that the magnitude of $\omega_1, \omega_2, \ldots, \omega_{2n}$ is the same. Combining the latter formula and the lemma, it can be concluded that $\omega_1, \omega_2, \ldots, \omega_{2n}$ are the vertices of a regular $2n$-gon centered at the origin or they are all 0. That is, $z_1, z_2, \ldots, z_{2n}$ form the vertices of a regular $2n$-gon in a certain order or they are all the same. In other words, $(x_1, y_1), \ldots, (x_{2n}, y_{2n})$ form the vertices of a regular $2n$-gon in a certain order or they are all the same.

Second Day
2nd March 2023

4. Given an acute-angled $\triangle ABC$, let H and O be its orthocenter and circumcenter, respectively. Let K be the midpoint of segment AH and that l be a line through O. Let P and Q be the orthogonal projections of B and C on l, respectively.

Prove that $KP + KQ \geqslant BC$.

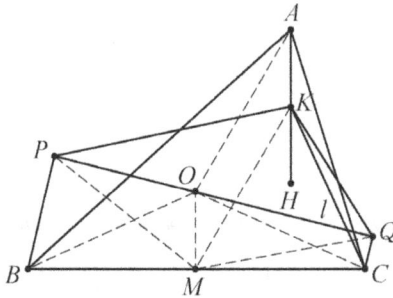

Fig. 8.1

Solution 1   As shown in Fig. 8.1, take the midpoint M be of BC. Connect $OM, KM, PM, QM, AO, BO, CO$, and then there is

$$AO = BO = CO.$$

Since $OM \perp BC, BP \perp PQ, CQ \perp PQ, \angle QPB = \angle OMC = \angle PQC = 90°$, points P, B, M, O are concyclic and points Q, C, M, O are concyclic.

Thus, $\angle QPM = \angle OBC = \angle OCB = \angle PQM$, and hence $\triangle MPQ \backsim \triangle OBC$, so

$$\frac{PQ}{PM} = \frac{BC}{OB}.$$

And by $OM \perp BC, AH \perp BC, AH = 2OM$, we know $AK \parallel OM$ and $AK = OM$. Thus, quadrilateral $OMKA$ is a parallelogram, so $AO = KM$.

In quadrilateral $KPMQ$, by Ptolemy's theorem, we have

$$KM \cdot PQ \leqslant KP \cdot QM + KQ \cdot PM.$$

Therefore,

$$KP + KQ \geqslant KM \cdot \frac{PQ}{PM} = AO \cdot \frac{BC}{OB} = BC.$$

The proof is completed.

Solution 2 Establish a complex coordinate system with O as the origin and l as the real axis, and use the lowercase alphabet to represent the complex coordinates of each point. We might as well let $|a| = |b| = |c| = 1$. Obviously, there is

$$k = a + \frac{1}{2}(b + c), p = \frac{1}{2}\left(b + \frac{1}{b}\right), q = \frac{1}{2}\left(c + \frac{1}{c}\right).$$

Therefore,

$$|k - p| = \left|a + \frac{1}{2}\left(c - \frac{1}{b}\right)\right| = \frac{1}{2}|2ab + bc - 1|.$$

Similarly, there is $|k - q| = \frac{1}{2}|2ac + bc - 1|$.

Consequently,

$$|k - p| + |k - q| = \frac{1}{2}|2ab + bc - 1| + \frac{1}{2}|2ac + bc - 1|$$

$$\geqslant \frac{1}{2}|2a(b - c)| = |b - c|.$$

The proof is completed.

5   Let $P(x), Q(x), R(x)$ and $S(x)$ be non-constant polynomials with real coefficients such that $P(Q(x)) = R(S(x))$. Suppose that the degree of $P(x)$ is a multiple of the degree of $R(x)$.

    Prove that there is a polynomial $T(x)$ with real coefficients such that

$$P(x) = R(T(x)).$$

Solution Assume there exists a polynomial $T(x)$ such that $S(x) = T(Q(x))$. By the given conditions, we have

$$P(Q(x)) = R(T(Q(x))),$$

and then $P(x) - R(T(x))$ always 0 in the range of $Q(x)$. Since $Q(x)$ is non-constant, $P(x) - R(T(x))$ has an infinite number of zeros, and hence $P(x) = R(T(x))$.

Next, we prove that there exists a polynomial $T(x)$ such that $S(x) = T(Q(x))$.

Comparing degrees, we know that $q = \deg Q(x) \mid \deg S(x) = s$. Select polynomials $T(x)$ and $M(x)$ such that $S(x) = T(Q(x)) + M(x)$ and $M(x)$ has the smallest degree.

If $M(x) \neq 0$, then $q = \deg Q$ does not divide $\deg M = m$. Otherwise there exist β and polynomial $M_1(x)$ such that $M(x) = \beta Q(x)^{\frac{m}{q}} + M_1(x)$. Let $T_1(x) = T(x) + \beta x^{\frac{m}{q}}$, and then $S(x) = T_1(Q(x)) + M_1(x)$, $\deg M_1 < \deg M_2$, a contradiction.

Denote α as the leading coefficient of $R(x)$, and $r = \deg R(x)$. Since $P(Q(x)) = R(S(x))$,

$$R(T(Q(x)) + M(x)) - R(T(Q(x))) = P(Q(x)) - R(T(Q(x)))$$

can be regarded as a polynomial in $Q(x)$. But the left side of the equation is the sum of $\alpha T(Q(x))^{r-1} M(x)$ and some low-order terms, and its degree is $\alpha s(r-1) + m$, which is not a multiple of q, a contradiction. Therefore, $M(x) = 0$, and then $S(x) = T(Q(x))$. The proof is completed.

6 Let r, g, b be non-negative integers. Let Γ be a connected graph on $r + g + b + 1$ vertices. The edges of Γ are each colored red, green or blue. It turns out that Γ has

(i) a spanning tree in which exactly r of the edges are red;
(ii) a spanning tree in which exactly g of the edges are green;
(iii) a spanning tree in which exactly b of the edges are blue.

Prove that Γ has a spanning tree in which exactly r of the edges are red, exactly g of the edges are green and exactly b of the edges are blue.

Note: A *spanning tree* of Γ is a graph which has the same vertices as Γ, with edges which are also edges of Γ, for which there is exactly one path between each pair of different vertices.

Solution We use mathematical induction on the number of vertices of Γ.

When $n = 2$, the conclusion is obviously true.

Assume that the conclusion holds for all $n \leqslant t$. Then when $n = t + 1$, we have $r + g + b = t$.

(1) If the vertices of graph Γ can be divided into two parts X and Y, such that the edges of the vertices in X and Y are colored the same color, respectively, we might as well be set to red. Then there is no spanning tree without red edges. Hence, $r \geqslant 1$.

According to the given condition, there is a spanning tree with exactly r red edges in Γ. Suppose one of the red edges is labeled as l, connecting two vertices that are located in X and Y, respectively. Remove the two vertices of l, consider this edge as a single vertex, and connect the edges originally connected to one of the vertices of l to l, obtaining a new graph Γ_0 with vertex number $t-1$. The spanning tree that has exactly r red edges, becomes a spanning tree with exactly $r-1$ red edges in Γ_0. There exists a spanning tree with exactly g green edges in Γ. If this tree contains l, then it becomes a spanning tree with exactly g green edges in Γ_0; if it does not contain l, then in Γ_0 the tree becomes a connected graph with exactly g green edges containing a cycle that contains the vertices originally in X and Y, respectively. Therefore, it contains at least one red edge, and after removing this red edge the graph becomes a tree with exactly g green edges.

Similarly, there exists a tree with exactly b blue edges in Γ_0. By the inductive hypothesis, there exists a spanning tree with $r-1$ red edges, g green edges, and b blue edges. Then, convert vertex l back into edge l. Reconnect the edges originally connected to l according to their colors to one of the two vertices of l, resulting in a tree with exactly r red edges, g green edges, and r blue edges. This is consistent with the conclusion.

(2) If the edges between any two parts of the vertices of graph Γ have at least two colors, consider the set of all graphs consisting of r red edges, g green edges, and b blue edges in Γ, none of which form cycles. Let the least number of branches in it be K. If K is not a tree, then there exists a branch in K containing a cycle that has at least two colors, and the vertices of this branch are connected to other vertices by edges of at least two colors. Without loss of generality, assume that all of them have a red color. Remove a red edge from the cycle, and connect another red edge to the branch. new graph. The new graph formed is still in the set and the number of branches is reduced, contradicting the assumption of minimality. Thus, K is a tree, and the conclusion holds.

Therefore, for any number of vertices n, the conclusion holds.

9

The 12th European Girls' Mathematical Olympiad

2023 (Slovenia)

The 12th European Girls' Mathematical Olympiad (EGMO) was held in Portorož, Slovenia from 13th to 19th April 2023. The European Girls' Mathematical Olympiad is a high-level international mathematics competition for female students. It aims to support and promote the popularity and development of mathematics among women, allowing more girls to experience the joy of mathematics, appreciate its beauty, and enjoy the pleasure of mathematical exploration. The Chinese Mathematical Society also hopes to use this competition as an opportunity to further promote mathematical education, research, and academic construction, and to nurture a new generation of young scientific and technological talents with international competitiveness. To this end, the Chinese Mathematical Society accepted the invitation to send representatives to participate in the competition for the first time, creating opportunities and conditions for improving China's ability and level of mathematics teaching and strengthening international competitions and exchanges, as well as providing a brand-new international stage for Chinese female students to demonstrate their basic mathematical skills and mathematical thinking.

A total of 213 students from 54 countries and regions (including 16 non-EU countries) participated in the competition. Through a week-long competition and activities, representatives competed with each other and strived for excellence in the field of mathematics competitions. This competition also invited Maryna Viazovska, a female Fields Medal winner who is renowned in the field of mathematics, to communicate with students from various countries.

The invited Chinese Girls' Olympic Team was led by Prof. Xiao Liang of Peking University, with Ms. Yu Dandan of the High School Affiliated to Renmin University of China as the deputy leader. The four members are Hu Shuwen and Yuan Lai from the High School Affiliated to Renmin University of China, Su Sihan from the No. 2 Middle School of Shijiazhuang, and Qiu Yiran from the Experimental High School Attached to Beijing Normal University.

In this competition, all members of the Chinese Girls' Olympic Team achieved full scores and gold medals, and it is the only team where every member achieved full scores and gold medal, winning the first place in the overall team rankings. The United States team and the Australian team won second and third places, respectively.

The teams that ranked in the top 10 in the overall team scores were:

1st place	China	168 points
2nd place	USA	164 points
3rd place	Australia	155 points
4th place	Ukraine	152 points
5th place	Turkey	150 points
6th place	Romania	144 points
7th place	Germany	141 points
8th place	Bulgaria	136 points
9th place	Hungary	132 points
10th place	Poland	113 points

First Day
(15th April 2023)

1 There are $n \geqslant 3$ positive real numbers $a_1, a_2, \ldots, a_n$. For each $i = 1, 2, \ldots, n$ we let $b_i = \frac{a_{i-1} + a_{i+1}}{a_i}$ (here we define a_0 to be a_n and a_{n+1} to be a_1). Assume that for all $i, j \in \{1, 2, \ldots, n\}$, there is $a_i \leqslant a_j$ if and only if $b_i \leqslant b_j$.

Prove that $a_1 = a_2 = \cdots = a_n$.

Solution 1 Without loss of generality, let a_s be the largest among $a_1, a_2, \ldots, a_n$, and a_t be the smallest among them. Then

$$b_s = \frac{a_{s-1} + a_{s+1}}{a_s} \leqslant 2 \leqslant \frac{a_{t-1} + a_{t+1}}{a_t} = b_t.$$

According to the condition, $b_t \leqslant b_s$ implies $a_t \geqslant a_s$. But a_t is the smallest and a_s is the largest. Therefore, all a_i are equal.

Solution 2   Assume all a_i are not all equal. Take an a_s that is the largest satisfying a_{s-1} and a_{s+1} are not all the largest. Therefore,

$$b_s = \frac{a_{s+1} + a_{s-1}}{a_s} < 2.$$

Since a_s is the largest, b_s is also the largest. So all $b_i < 2 (1 \leqslant i \leqslant n)$.

However, on the other hand,

$$\begin{aligned} b_1 b_2 \cdots b_n &= \frac{a_n + a_2}{a_1} \cdot \frac{a_1 + a_3}{a_2} \cdots \cdots \frac{a_{n-1} + a_1}{a_n} \\ &\geqslant \frac{2\sqrt{a_n a_2}}{a_1} \cdot \frac{2\sqrt{a_1 a_3}}{a_2} \cdots \cdots \frac{2\sqrt{a_{n-1} a_1}}{a_n} = 2^n, \end{aligned}$$

which contradicts all $b_i < 2$.

② We are given an acute-angled triangle ABC. Let D be the point on its circumcircle such that AD is a diameter. Suppose that points K and L lie on segments AB and AC, respectively, such that DK and DL are tangent to the circumcircle of triangle AKL.

Prove that line KL passes through the orthocenter of triangle ABC.

Note: The altitudes of a triangle meet at its orthocenter.

Solution   As shown in Fig. 9.1, let M be the midpoint of KL. Since DK and DL are tangent to $\odot AKL$, we have $DK = DL$, and hence $DM \perp KL$. And since AD is the diameter, we know $DB \perp AB$ and $DC \perp AC$. Therefore, B, D, M, K are concyclic, and C, D, M, L are concyclic.

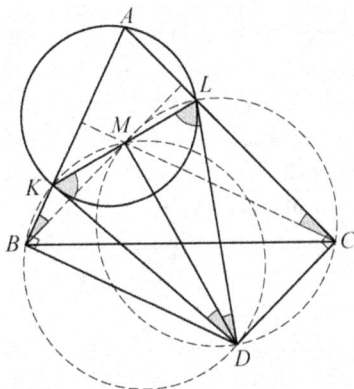

Fig. 9.1

By using the properties of concyclic points and the alternate angles, we have

$$\angle MBK = \angle MDK = 90^\circ - \angle MKD$$

$$= 90^\circ - \angle LKD = 90^\circ - \angle A,$$

$$\angle MCL = \angle MDL = 90^\circ - \angle MLD$$

$$= 90^\circ - \angle KLD = 90^\circ - \angle A.$$

From this we know $BM \perp AC$ and $CM \perp AB$. Therefore, line KL passes through the orthocenter of triangle ABC.

3 Let k be a positive integer. Lexi has a dictionary $\mathcal{D}$ consisting of some k-letter strings containing only the letters A and B. Lexi would like to write either the letter A or the letter B in each cell of a $k \times k$ grid so that each column contains a string from $\mathcal{D}$ when read from top-to-bottom and each row contains a string from $\mathcal{D}$ when read from left-to-right.

What is the smallest integer m such that if $\mathcal{D}$ contains at least m different strings, then Lexi can fill her grid in this manner, no matter what strings are in $\mathcal{D}$?

Solution We claim that the smallest positive integer m is 2^{k-1}.

First, we show that $m = 2^{k-1} - 1$ does not satisfy the conditions of the problem. Let $\mathcal{D}$ be all strings of length k that start with A but are not entirely A. There are $2^{k-1} - 1$ such strings in total. If Lexi can fill out the grid, then the first row must be chosen from $\mathcal{D}$, so there must be letter B. But the column where the letter B is located is not a string that appears in $\mathcal{D}$. This is a contradiction.

Next, we demonstrate that $m = 2^{k-1}$ satisfy the conditions of the problem. First note that if $\mathcal{D}$ contains the strings all A or all B, we just fill the grid entirely with A or entirely with B. Hereafter, we assume that $\mathcal{D}$ does not contain the strings of all A or all B.

For a string L of length k, let $\bar{L}$ represent the string obtained by changing all A's to B's and all B's to A's in L. We call this the conjugate of L. By the pigeonhole principle, when $|\mathcal{D}| \geqslant 2^{k-1}$ and $\mathcal{D}$ does not contain the strings all A and all B, $\mathcal{D}$ must contain some string L and its conjugate $\bar{L}$. At this point, the $k \times k$ grid constructed by Lexi is as follows: if the ith letter of L is A, fill in the ith row of the $k \times k$ gird with L; if the ith letter of L is B, fill in the ith row of the $k \times k$ gird with $\bar{L}$. Note that this construction also ensures: if the ith letter of L is A, then the ith column from top-to-bottom is L; if the ith letter of L is B, then the ith column from top-to-bottom is $\bar{L}$. This filling method satisfies the conditions.

Second Day
(16th April 2023)

4   Turbo the snail sits on a point on a circle with circumference 1. Given an infinite sequence of positive real numbers $c_1, c_2, c_3, \ldots$, Turbo successively crawls distances $c_1, c_2, c_3, \ldots$ around the circle, each time choosing to crawl either clockwise or anticlockwise.

For example, if the sequence $c_1, c_2, c_3, \ldots$ is $0.4, 0.6, 0.3, \ldots$, then Turbo may start crawling as follows (Fig. 9.2):

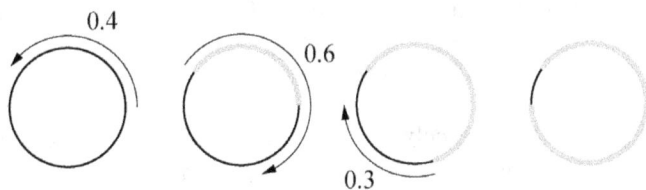

Fig. 9.2

Determine the largest constant $C > 0$ with the following property: for every sequence of positive real numbers $c_1, c_2, c_3, \ldots$ with $c_i < C$ for all i, Turbo can (after studying the sequence) ensure that there is some point on the circle that it will never visit or crawl across.

Solution   The desired positive real number is $C = \frac{1}{2}$.

First, we will show that $C = \frac{1}{2}$ satisfies the requirement of the question. Suppose Turbo starts at point P and chooses any point $Q \neq P$ on the circle. Turbo will always have a strategy to avoid reaching Q because there are two different arcs from P to Q (clockwise and anticlockwise), one of which has a length of at least $\frac{1}{2}$. Turbo can always choose to move in the direction and thus it will never reach or cross Q, for each $c_i = \frac{1}{2}$.

Next, we demonstrate that any constant $C > \frac{1}{2}$ does not meet the requirement. Denote $\varepsilon = C - \frac{1}{2} > 0$. Take the sequence $\frac{1}{2}, \frac{1+\varepsilon}{2}, \frac{1}{2}, \frac{1+\varepsilon}{2}, \ldots$ That is, $c_{2i-1} = \frac{1}{2}, c_{2i} = \frac{1+\varepsilon}{2} (i \in \mathbf{N})$. Note that if Turbo continues to crawl in the same direction twice in a row, it will pass all points on the circle because the sum of any two consecutive terms in c_i is greater than 1. Assume this does not occur. Then Turbo must change direction after each move. After every two moves, Turbo will move $\frac{\varepsilon}{2}$ along a fixed direction. After enough moves, Turbo will traverse every point on the circle. Hence, $C > \frac{1}{2}$ does not meet the requirement.

⑤ We are given a positive integer $s \geqslant 2$. For each positive integer k, we define its twist k' as follows: write k as $as + b$, where a, b are non-negative integers and $b < s$, and then $k' = bs + a$. For the positive integer n, consider the infinite sequence $d_1, d_2, \ldots$ where $d_1 = n$ and d_{i+1} is the twist of d_i for each positive integer i.

Prove that this sequence contains 1 if and only if the remainder when n is divided by $s^2 - 1$ is either 1 or s.

Solution First, note that for any i, if we denote $d_i = as + b$, then

$$d_{i+1} - sd_i = (bs + a) - s(as + b) = a(1 - s^2)$$

is a multiple of $s^2 - 1$, namely, $d_{i+1} \equiv sd_i \bmod (s^2 - 1)$. Therefore, if some d_n is congruent to 1 or s modulo $s^2 - 1$, then $s^{n-1}d_1 \equiv 1$ or s modulo $s^2 - 1$. Hence, $d_1 \equiv 1$ or s modulo $s^2 - 1$.

Conversely, if $d_1 \equiv 1$ or s modulo $s^2 - 1$, then every $d_n \equiv s^{n-1}$ or s^n modulo $s^2 - 1$, i.e., $d_n \equiv 1, s \bmod (s^2 - 1)$. Therefore,

$$d_{i+1} - d_i = (bs + a) - (as + b) = (b - a)(s - 1) > 0.$$

And if $d_i \leqslant s^2 - 1$, then $a \leqslant s - 1$, so $d_{i+1} \leqslant s^2 - 1$. In summary, the sequence $d_1, d_2, \ldots$ will start to strictly decrease until it is less than or equal to $s^2 - 1$, and then it will always remain less than or equal to $s^2 - 1$. Combining this with $d_i \equiv 1$ or s modulo $s^2 - 1$, we know that d_i alternates between 1 and s when i is sufficiently large.

⑥ Let ABC be a triangle with circumcircle Ω. Let S_b and S_c respectively denote the midpoints of the arcs $\overset{\frown}{AC}$ and $\overset{\frown}{AB}$ that do not contain the third vertex. Let N_a denote the midpoint of arc $\overset{\frown}{BAC}$ (the arc $\overset{\frown}{BC}$ containing A). Let I be the incenter of ABC. Let ω_b be the circle that is tangent to AB and internally tangent to Ω at S_b, and let ω_c be the circle that is tangent to AC and internally tangent to Ω at S_c. Show that line IN_a, and the line through the intersections of ω_b and ω_c meet on Ω.

(The incenter of a triangle is the center of its incircle, the circle inside the triangle that is tangent to all three sides.)

Solution First, we will show that point A lies on the radical axes of ω_b and ω_c. This step does not require point N_a.

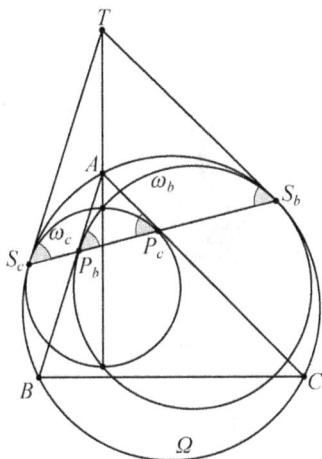

Fig. 9.3

As shown in Fig. 9.3, let the tangent lines to Ω at points S_b and S_c intersect at point T, which is the radical center of ω_b, ω_c and Ω. Let AB and S_bS_c intersect at point P_b, and let AC and S_bS_c intersect at point P_c. Then triangle TS_cS_b is an isosceles triangle. And since $AB \parallel TS_c$ and $AC \parallel TS_b$, there is

$$\angle AP_bP_c = \angle TS_cS_b = \angle S_cS_bT = \angle P_bP_cA.$$

From these we can see that ω_b passes through point P_b, ω_c passes through point P_c, and AP_b and AP_c are exactly tangents to ω_b and ω_c. In particular, A lies on the radical axes of ω_b and ω_c. Therefore, the line passing through the intersection of ω_b and ω_c is the line AT.

Next we show that line TA and line IN_a intersect at a point on circle Ω. Let line TA intersect circle Ω at point X. Let line XI intersect circle Ω at point N. We need to show that $N = N_a$.

As shown in Fig. 9.4, let S_a be the midpoint of arc $\overarc{BC}$ (not containing point A).

Note that the triples of points A, I, S_a; B, I, S_b and C, I, S_c are collinear because they lie on the corresponding angle bisectors, respectively. And

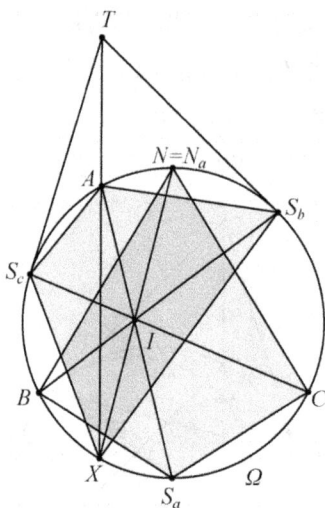

Fig. 9.4

note that AS_cXS_b is a harmonic quadrilateral, because the tangents at S_b and S_c and line AX intersect at point T. By "inverting" it with respect to point I, this harmonic quadrilateral becomes harmonic quadrilateral S_aCNB. (This is an easy property to prove for a harmonic quadrilateral:

$$\frac{AS_c}{S_aC} \cdot \frac{XS_b}{NB} \cdot \frac{S_aB}{AS_b} \cdot \frac{NC}{XS_c} = \frac{AI}{IC} \cdot \frac{IX}{IB} \cdot \frac{IB}{IA} \cdot \frac{IC}{IX} = 1,$$

and thus $AS_c \cdot XS_b = AS_b \cdot XS$ implies $S_aC \cdot NB = S_aB \cdot NC$.) However, $S_aB = S_aC$.

Therefore, $NB = NC$. It follows that $N = N_a$.

10

International Mathematical Olympiad

2023 (Chiba, Japan)

The 64th International Mathematical Olympiad (IMO) was held from 2nd July to 13th July 2023 in Chiba, Japan. A total of 612 contestants from 112 countries or regions participated the competition. After two days of exams on the 8th and 9th, all six members of the Chinese team won gold medals, including Wang Chunji and Shi Haojia with full scores. The Chinese team won the first place in the team for the fifth consecutive year.

The Chinese team:

Leader: Xiao Liang (Peking University)

Deputy leader: Qu Zhenhua (East China Normal University)

Observers: Fu Yunhao (Southern University of Science and Technology), Wang Bin (Academy of Mathematics and Systems Science, Chinese Academy of Sciences)

Head coach: Xiong Bin (East China Normal University)

Team members:

Wang Chunji (Shanghai High School), 10th grade, 42 marks, gold medalist

Shi Haojia (Zhuji Hailiang Senior High School), 10th grade, 42 marks, gold medalist

Liang Xingjian (The High School Attached to Hunan Normal University), 12th grade, 41 marks, gold medalist

Zhang Xinliang (Zhenhai High School of Ningbo), 12th grade, 41 marks, gold medalist

Sun Qi'ao (Shanghai High School), 11th grade, 35 marks, gold medalist

Jiang Zhicheng (Shenzhen Middle School), 11th grade, 42 marks, gold medalist

The top 10 teams in the total scores:

1.	China	240 marks
2.	United States of America	222 marks
3.	South Korea	215 marks
4.	Romania	208 marks
5.	Canada	183 marks
6.	Japan	181 marks
7.	Vietnam	180 marks
8.	Turkey	176 marks
9.	India	174 marks
10.	Chinese Taipei	173 marks

Gold medal: score $\geqslant 32$ marks;

Silver medal: score $\geqslant 25$ marks;

Bronze medal: score $\geqslant 18$ marks.

Between May 8th and 17th, the Chinese team received first phase of training at the High School Attached to Hunan Normal University, Changsha. The coaches Jin Chunlai, Yu Jun, Fu Yunhao, Leng Fusheng, Wu Hao and Xiao Liang gave lectures and tests. From 1st June to 8th June, the Chinese team received second phase of training at Shanghai High School, Shanghai. The coaches Lu Sheng, Leng Gangsong, Qu Zhenhua, Zhang Sihui and Ai Yinghua gave lectures and two tests.

From 27th June, the Chinese team received pre-departure training and preparation at Peking University, Beijing. Academician Tian Gang, President of Chinese Mathematical Society; Professor Gong Fuzhou, Secretary-General of the Chinese Mathematical Society; Professor Chen Dayue, Dean of School of Mathematical Sciences, Peking University; Professor Sun Zhaojun, Associate Dean; Professor Liu Bin; coaches Xu Disheng, Wang Bin, Fu Yunhao, Qu Zhenhua and Yao Yijun, and gold medalists of IMO 2022, Zhang Zhicheng and Jiang Cheng, gave reports, tutoring or had discussions with the team members. Accommodation and other related work was provided by the School of Mathematical Sciences, Peking University and Beijing International Centre for Mathematical Research.

First Day
(8th July 2023)

Problem 1 Determine all composite numbers $n > 1$ that satisfy: if $d_1, d_2, \ldots, d_k$ are all the positive divisors of n, $1 = d_1 < d_2 < \cdots < d_k = n$, then d_i divides $d_{i+1} + d_{i+2}$ for every $1 \leqslant i \leqslant k - 2$.

(Contributed by Colombia)

Solution The desired composite numbers have the form $n = p^a$ where p is a prime and $a \geqslant 2$ is an integer. Clearly, if $n = p^a (a \geqslant 2)$, then $d_i = p^{i-1}(i = 1, 2, \ldots, a + 1)$ and $d_i = p^{i-1}$ divides $d_{i+1} + d_{i+2} = p^i(1 + p)$ for all $i = 1, 2, \ldots, a - 1$.

Now suppose n has at least two distinct prime factors, p being the smallest and q being the second smallest, then for some $i \in \mathbf{N}_+$, we have

$$d_k = n, d_{k-1} = \frac{n}{p}, \ldots, d_{k-i} = \frac{n}{p^i}, d_{k-i-1} = \frac{n}{q}.$$

It is known that

$$d_{k-i-1} \left| (d_{k-i} + d_{k-i+1}) \Rightarrow \frac{n}{q} \right| \left(\frac{n}{p^i} + \frac{n}{p^{i-1}} \right) = \frac{n}{p^i}(p+1),$$

however, $v_p\left(\frac{n}{q}\right) > v_p\left(\frac{n}{p^i}\right)$, and thus the above divisibility relation cannot hold. This implies that n cannot have two or more prime factors; n must be a prime power.

Problem 2 Let ABC be an acute-angled triangle with $AB < AC$. Let Ω be the circumcircle of ABC. Let S be the midpoint of the arc CB of Ω containing A. The perpendicular from A to BC meets BS at point D and meets Ω again at point $E \neq A$. The line through D parallel to BC meets line BE at point L. Denote the circumcircle of triangle BDL by ω. Let ω meets Ω again at point $P \neq B$. Prove that the line tangent to ω at P meets line BS on the internal angle bisector of $\angle BAC$.

(Contributed by Portugal)

Solution 1 As shown in Fig. 10.1, let S' be the midpoint of the minor arc $\overset{\frown}{BC}$ of ω. Then SS' is a diameter of Ω and AS' is the bisector of $\angle BAC$. Let the tangent line of ω at P meet Ω again at point $Q(Q \neq P)$. Then $\angle SQS' = 90°$. We will prove the corresponding sides of $\triangle APD$ and $\triangle S'QS$ are parallel, which implies that the lines connecting the corresponding

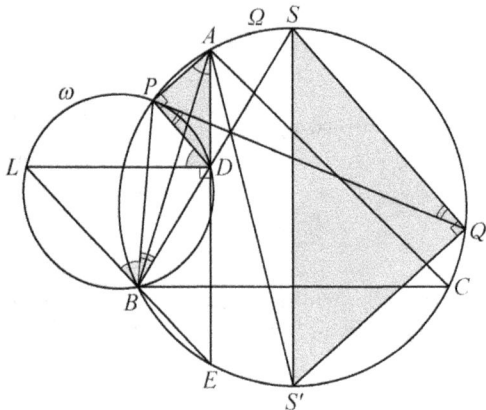

Fig. 10.1

vertices, namely the bisector of $\angle BAC$, the tangent line of ω at P, and DS are concurrent (the intersection point is the internal homothetic center).

First, as A, P, B, E are concyclic and D, P, L, B are concyclic, we find

$$\angle PAD = \angle PAE = 180° - \angle EBP = \angle PBL = \angle PDL = 90° - \angle ADP,$$

and thus $AP \perp DP$. Now we see that

- The line segments ADE and SS' are both perpendicular to BC, and so $AD \parallel S'S$.
- The segment PQ touches ω at P, implying that $\angle DPQ = \angle DBP = \angle SBP = \angle SQP$ and thus $PD \parallel QS$.
- Since $AP \perp PD, PD \parallel QS, QS \perp S'Q$, we have $AP \parallel S'Q$, and the corresponding sides of $\triangle APD$ and $\triangle S'QS$ are parallel.

This completes the proof.

Solution 2   As illustrated in Fig. 10.2, let S' be the midpoint of the minor arc $\overparen{BC}$ of ω, S' is the antipode of S, and thus $AES'S$ is an isosceles trapezium, $\angle S'BS = \angle S'PS = 90°$. Let segments AE and PS' meet at point T; lines AP and $S'B$ meet at point M. We shall Prove that L, P, and S are collinear; T and M lie on ω.

- Since $\angle LPB = \angle LDB = 90° - \angle BDE = 90° - \angle BSS' = \angle SS'B = 180° - \angle BPS$, the points L, P, S are indeed collinear.
- Since SS' is a diameter of Ω, the lines LPS and PTS' are orthogonal. We also have $LD \parallel BC, BC \perp AE$. Hence, $\angle LDT = \angle LPT = 90°, T \in \omega$.

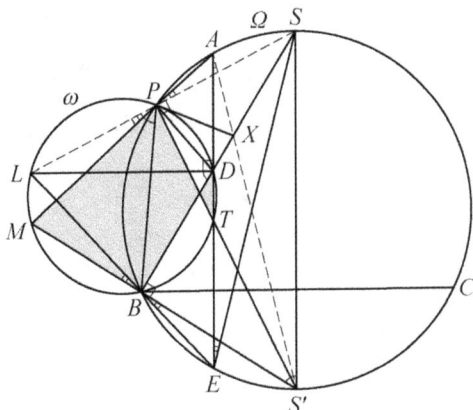

Fig. 10.2

- Since $\angle LPM = \angle SPA = \angle SEA = \angle EAS' = \angle EBS' = \angle LBM$, the points $M, B, P,$ and L are concyclic, and $M \in \omega$.

Now, let X be the point of intersection of line BDS and the tangent line of ω at P. Apply Pascal's theorem to the degenerate cyclic hexagon $PPMBDT$ to find that $PP \cap BD = X, PM \cap DT = A$ and $MB \cap TP = S'$ are collinear. Therefore, X lies on line AS', that is, the bisector of $\angle BAC$.

Problem 3 For each integer $k \geqslant 2$, determine all infinite sequences of positive integers $a_1, a_2, \ldots$ for which there exists a polynomial P of the form $P(x) = x^k + c_{k-1}x^{k-1} + \cdots + c_1 x + c_0$, where $c_0, c_1, \ldots, c_{k-1}$ are nonnegative integers, such that

$$P(a_n) = a_{n+1}a_{n+2} \cdots a_{n+k} \qquad (*)$$

for every integer $n \geqslant 1$.

(Contributed by Malaysia)

Solution 1 The infinite sequences that meet the problem conditions are those arithmetic sequences $a_1, a_2, \ldots$ with positive initial terms and nonnegative common differences: $a_n = a + nd$ where $a, n \in \mathbf{N}_+, d \in \mathbf{N}$. First, if $a_n = a + nd$, taking $P(x) = (x+d)(x+2d)\cdots(x+kd)$, it is not difficult to check that $(*)$ holds for every integer $n \geqslant 1$.

Next, we prove necessity in two steps.

Step 1: The sequence $\{a_n\}$ is either a constant sequence or strictly increasing.

Suppose the assertion is false: $a_{n_0-1} \geqslant a_{n_0}$ for some $n_0 \geqslant 2$. Then

$$a_{n_0} a_{n_0+1} \cdots a_{n_0+k-1} = P(a_{n_0-1}) \geqslant P(a_{n_0}) = a_{n_0+1} a_{n_0+2} \cdots a_{n_0+k},$$

yielding $a_{n_0} \geqslant a_{n_0+k}$. Let a_{n_1} be the first term after a_{n_0} and not exceeding a_{n_0}. We have $n_1 \leqslant n_0 + k$ and $a_{n_1-1} \geqslant a_{n_1}$.

Similarly, we can find $a_{n_2-1} \leqslant a_{n_2}$ with subscript $n_2 \in (n_1, n_1 + k]$. Continue this process to obtain $a_{n_0} \geqslant a_{n_1} \geqslant a_{n_2} \geqslant \cdots$. Now, all $P(a_{n_i})$ are bounded; as $n_i - n_{i-1} \leqslant k$, all a_j are bounded as well. (In fact, for each a_j, there exists n_i such that $j \in [n_i+1, n_i+k]$, and so $a_j \leqslant P(a_{n_i})$.) Let a_M be the largest element of $\{a_n\}$. We have $P(a_M) = a_{M+1} a_{M+2} \cdots a_{M+k} \leqslant a_M^k$, which requires $P(x) = x^k$ and $a_{M+1} = a_{M+2} = \cdots = a_{M+k}$. By induction, it is easy to see that $\{a_n\}$ is a constant sequence.

Step 2: If $\{a_n\}$ is strictly increasing, then it is an arithmetic sequence with positive common difference.

Let $P(x) = x^k + d_{k-1} x^{k-1} + \cdots + d_0$. Define $C = d_0 + \cdots + d_{k-1} + 1$. For any $m \in \mathbf{N}_+$, we have $P(m) \leqslant m^k + C m^{k-1}$. Note that

$$a_{n+1} a_{n+2} \cdots a_{n+k} = P(a_n) \leqslant a_n^k + C a_n^{k-1}.$$

Since each $a_{n+j} > a_n$, it follows that $a_{n+i} \leqslant a_n + C$ for each $i = 1, \ldots, k$.

If $a_n > C^k \cdot 2^k$ (obviously $a_n > \max\{d_0, \ldots, d_{k-1}\}$), then for $i = 1, \ldots, k$, $a_{n+i} = a_n + \delta_i, \delta_i \in (0, C)$. Note that

$$
\begin{aligned}
a_n^k + d_{k-1} a_n^{k-1} + \cdots + d_0 &= P(a_n) = a_{n+1} a_{n+2} \cdots a_{n+k} \\
&= (a_n + \delta_1)(a_n + \delta_2) \cdots (a_n + \delta_k) \\
&= a_n^k + (\delta_1 + \cdots + \delta_k) a_n^{k-1} + \cdots + \delta_1 \cdots \delta_k.
\end{aligned}
$$

We infer that d_i equals the elementary symmetric polynomial in $\delta_1, \ldots, \delta_k$ of degree $k - i$. (A simple viewpoint is to consider both sides as base-a_n numbers.) This means

$$P(x) = (x + \delta_1) \cdots (x + \delta_k),$$

in particular, all δ_i's are independent of n. Consequently, if $a_m > C^k \cdot 2^k$, then $a_{m+i} = a_m + \delta_i$ for each $i = 1, \ldots, k$. Likewise for a_{m+1}, we have $a_{m+i} = a_m + \delta_{i-1} + \delta_1$, and thus $\delta_i = i\delta_1$. It follows that for sufficiently large n, $a_n, a_{n+1}, \ldots$ form an arithmetic sequence.

Using backward induction, we can easily verify the entire sequence $\{a_n\}$ is an arithmetic sequence.

Solution 2 (Edited from Shi Jiahao and Jiang Zhicheng's solution) The proof that all arithmetic sequences $\{a_n\}$ with positive initial terms and nonnegative common differences meet the problem conditions is the same as in Solution 1. We show necessity as follows.

(0) First, if $P(x) = x^k$, taking a_M to be the smallest element of $\{a_n\}$, then $P(a_M) = a_{M+1} \cdots a_{M+k} \geqslant a_M^k$ in which equality holds, and we get $a_M = a_{M+1} = \cdots = a_{M+k}$. From the recurrence, $a_n = a_M$ for every n.

Next, assume $P(x) \neq x^k$. Denote $M = c_{k-1} + c_{k-2} + \cdots + c_0 + 2$. We prove the statement in four steps.

(1) **Claim:** for any positive integer N, from some term on, every term in the sequence is larger than N.

As $a_{n+1} \cdots a_{n+k} = P(a_n) > a_n^k$, we have $\max\{a_{n+1}, \ldots, a_{n+k}\} > a_n$ and

$$\max\{a_{n+k+1}, \ldots, a_{n+2k}\} > \max\{a_{n+1}, \ldots, a_{n+k}\}.$$

When $n > kMN^k$,

$$P(a_n) = a_{n+1} \cdots a_{n+k} \geqslant \max\{a_{n+1}, \ldots, a_{n+k}\} \geqslant \left\lfloor \frac{n}{k} \right\rfloor \geqslant MN^k,$$

and hence $a_n > N$.

(2) **Claim:** there is no pair (i, j) such that $|j - i| \leqslant k$ and $a_j - a_i > M$.

Otherwise, take a pair (i, j) such that a_i is minimal, and on this basis, a_j is maximal (the existence of maximal a_j is guaranteed by $|j - i| \leqslant k$ and (1); if it is not unique, pick any pair).

If $i < j$, then from $a_j > a_i + M$ and

$$a_i^{k-1}(a_i + M) > P(a_i) = \left(\prod_{i+1 \leqslant t \leqslant i+m, \ t \neq j} a_t \right) a_j,$$

we can find $i + 1 \leqslant t \leqslant i + k$ such that $a_t < a_i$, indicating that (t, j) is another pair, but a_i is assumed to be minimal, which is a contradiction.

If $i > j$, as $a_j^k < P(a_j) = a_{j+1} \cdots a_{j+k}$, there exists $j + 1 \leqslant t \leqslant j + k$ such that $a_t > a_j$, (i, t) is another pair, but a_j is assumed to be maximal, which is a contradiction.

(3) **Claim:** when n is sufficiently large, the polynomials $P(x)$ and $Q(x) = (x + a_{n+1} - a_n)(x + a_{n+2} - a_n) \cdots (x + a_{n+k} - a_n)$ are identical.

Note that in the expansion of $Q(x)$, the absolute value of each coefficient has an upper bound independent of n (since $|a_{n+i} - a_n| \leqslant M, 1 \leqslant i \leqslant k$).

This implies that for $Q(x) - P(x)$, a polynomial with integer coefficients and degree not exceeding k, the absolute values of the coefficients are all bounded. If $Q(x) - P(x)$ is not the zero polynomial, then the absolute values of its integral roots are also bounded (independent of n). When n is sufficiently large, (1) indicates that a_n is large, and $P(a_n) = Q(a_n)$ indicates that $P(x) = Q(x)$.

(4) **Claim:** $\{a_n\}$ is an arithmetic sequence.

Take n sufficiently large such that $P(x) = (x + d_1)(x + d_2)\cdots(x + d_k)$, where $d_i = a_{n+i} - a_n$. Since P has no positive roots, $d_1, d_2, \ldots, d_k \geqslant 0$, the sequence is non-decreasing when n is sufficiently large, and hence $d_1 \leqslant d_2 \leqslant \cdots \leqslant d_k$. Note that the unordered factorization of P into linear terms is unique. Therefore, for every n sufficiently large, $a_{n+i} - a_n = d_i$ $(1 \leqslant i \leqslant k)$, and in particular, $a_{n+1} - a_n = d_1$. It follows that for n sufficiently large, $a_n, a_{n+1}, \ldots, a_{n+k}$ is an arithmetic sequence with common difference d_1 and $d_i = id_1$ $(1 \leqslant i \leqslant k)$.

Suppose $a_1, a_2, \ldots$ is not an arithmetic sequence. Take the largest m such that $a_m \neq a_{m+1} - d_1$. Then

$$P(a_m) = a_{m+1}\cdots a_{m+k} = a_{m+1}(a_{m+1} + d_1)\cdots(a_{m+1} + (k-1)d_1)$$
$$= P(a_{m+1} - d_1),$$

but $P(x) = (x + d_1)\cdots(x + kd_1)$ is strictly increasing for $x \in [-d_1, \infty)$, a contradiction!

Second Day
(9th July 2023)

Problem 4   Let $x_1, x_2, \ldots, x_{2023}$ be pairwise different positive real numbers such that

$$a_n = \sqrt{(x_1 + x_2 + \cdots + x_n)\left(\frac{1}{x_1} + \frac{1}{x_2} + \cdots + \frac{1}{x_n}\right)}$$

is an integer for every $n = 1, 2, \ldots, 2023$. Prove that $a_{2023} \geqslant 3034$.

(Contributed by Holland)

Solution   It suffices to prove, for every $n \in \{1, \ldots, 2021\}$, $a_{n+2} \geqslant a_n + 3$. If this is true, then $a_{2023} \geqslant a_{2021} + 3 \geqslant a_{2019} + 6 \geqslant \cdots \geqslant a_1 + 3033 = 1 + 3033 = 3034$.

By Cauchy's inequality, we have

$$((x_1 + x_2 + \cdots + x_n) + (x_{n+1} + x_{n+2}))$$

$$\cdot \left(\left(\frac{1}{x_1} + \frac{1}{x_2} + \cdots + \frac{1}{x_n} \right) + \left(\frac{1}{x_{n+1}} + \frac{1}{x_{n+2}} \right) \right)$$

$$\geq \left(\left(\sqrt{(x_1 + x_2 + \cdots + x_n) \left(\frac{1}{x_1} + \frac{1}{x_2} + \cdots + \frac{1}{x_n} \right)} \right. \right.$$

$$\left. \left. + \sqrt{(x_{n+1} + x_{n+2}) \left(\frac{1}{x_{n+1}} + \frac{1}{x_{n+2}} \right)} \right) \right)^2 .$$

It follows that

$$a_{n+2} \geq a_n + \sqrt{(x_{n+1} + x_{n+2}) \left(\frac{1}{x_{n+1}} + \frac{1}{x_{n+2}} \right)} \geq a_n + 2.$$

Since x_i's are pairwise different, $x_{n+1} \neq x_{n+2}$, the second relation in the above cannot be equal. Since a_n's are integers, we reach $a_{n+2} \geq a_n + 3$ and the problem is proved.

Problem 5 Let n be a positive integer. A *Japanese triangle* consists of $1 + 2 + \cdots + n$ circles arranged in an equilateral triangular shape such that for each $i = 1, 2, \ldots, n$, the ith row contains exactly i circles, exactly one of which is colored red. A *ninja path* in a Japanese triangle is a sequence of n circles obtained by starting in the top row, then repeatedly going from a circle to one of the two circles immediately below it and finishing in the bottom row. Figure 10.3 shows an example of a Japanese triangle with $n = 6$, along with a ninja path in that triangle containing two red circles.

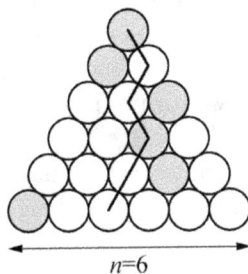

$n=6$

Fig. 10.3

In terms of n, find the greatest k such that in each Japanese triangle there is a ninja path containing at least k red circles.

(Contributed by Holland)

Solution   The largest integer k is $\lfloor \log_2 n \rfloor + 1$.

Let $N = \lfloor \log_2 n \rfloor$. So, $2^N \leqslant n \leqslant 2^{N+1} - 1$. We construct a Japanese triangle such that every ninja path contains at most $N + 1$ red circles. As shown in Fig. 10.4 ($n = 10$), in row $i, i = 2^a + b, 0 \leqslant a \leqslant N, 0 \leqslant b < 2^a$, the $(2b + 1)$th circle is colored red. It is not difficult to figure out that for $a = 0, 1, \ldots, N$, every ninja path passes through at most one red circle in rows $2^a, 2^a + 1, \ldots, 2^{a+1} - 1$. Hence, every ninja path passes through at most $N + 1$ red circles.

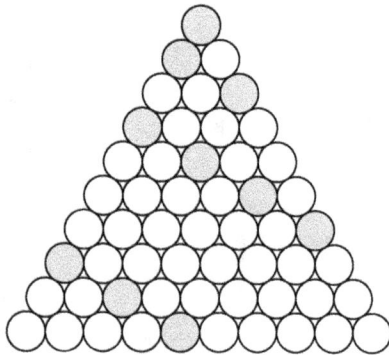

Fig. 10.4

Next, we present four proofs that in every Japanese triangle there is a ninja path traversing at least $N + 1$ red circles.

Proof 1   For convenience, we generalize the definition of ninja path to any contiguous subsequence so that a ninja path does not necessarily start from the top or finish at the bottom. Since a subsequence contains the same number or fewer red circles, we turn to prove the statement for generalized ninja paths.

For each circle C, assign to it the maximum number of red circles that a ninja path starting from the top and finishing at C can have. An example is in Fig. 10.5. Note that

- If circle C is not red, then the number in C is the maximum of the one or two numbers above it;

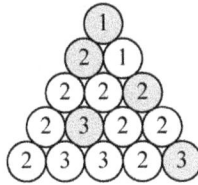

Fig. 10.5

- If circle C is red, then the number in C is the maximum of the number(s) above it plus 1.

Let the numbers in row i be $v_1, \ldots, v_i$ in which v_m is the largest. From the above observation, we deduce that the numbers in row $i + 1$ are at least $v_1, \ldots, v_{m-1}, v_m, v_m, v_{m+1}, \ldots, v_i$, without considering the red circle in row $i+1$. With the presence of the red circle, the number assigned to that circle increases by 1. Hence, the sum of the numbers in row $i + 1$ is at least

$$(v_1 + \cdots + v_i) + v_m + 1,$$

which leads us to the following result:

Lemma 1 *Let σ_k be the sum of the numbers in row k. For $0 \leqslant j \leqslant N$,*

$$\sigma_{2^j} \geqslant j \cdot 2^j + 1.$$

Proof of Lemma 1 We induct on j. When $j = 0$, the conclusion is obvious as the top circle always has the number 1. Suppose we have $\sigma_{2^j} \geqslant j \cdot 2^j + 1$, then in row 2^j the largest number is at least $j + 1$. For all $k \geqslant 2^j$, in row k there is a circle with number at least $j + 1$, and in the next row we have

$$\sigma_{k+1} \geqslant \sigma_k + (j + 1) + 1 = \sigma_k + (j + 2).$$

Therefore,

$$\sigma_{2^{j+1}} \geqslant \sigma_{2^j} + 2^j(j + 2) \geqslant j \cdot 2^j + 1 + 2^j(j + 2)$$
$$= (j + j + 2)2^j + 1 = (j + 1)2^{j+1} + 1,$$

completing the inductive proof.

For $j = N$, in row 2^N there is a circle whose number is at least $N + 1$. So, there exists a ninja path that passes through at least $N + 1$ red circles.

Proof 2 We generalize the definition of ninja path and assign a number to each circle as in proof 1. For each positive integer i, let e_i be the number of red circles with number i.

Lemma 2 *If the red circle in row l has number i, then $e_i \leqslant l$.*

Proof of Lemma 2 If two red circles are assigned the same number, then they cannot be connected by a ninja path. We divide the large triangle into a smaller one whose top is at the red circle in row l and $l-1$ lines as shown in Fig. 10.6. Evidently, these l subsets contain all the circles.

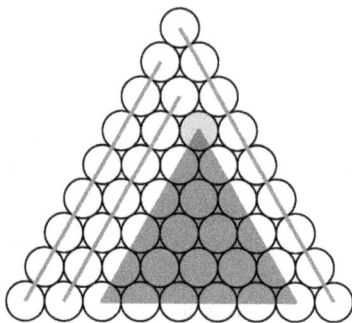

Fig. 10.6

In each subset, at most one red circle can have number i. Thus, $e_i \leqslant l$.

Note that if C is a red circle with number $i \geqslant 2$, then there must be another red circle above C's row and with number $i-1$.

Lemma 3 *For each positive integer i, we have $e_i \leqslant 2^{i-1}$.*

Proof of Lemma 3 We induct on i. The base case $i = 1$ is evident as the only red circle with number 1 is the top circle. Suppose the induction hypothesis is true for $1 \leqslant i \leqslant j-1$. If $e_j = 0$, the lemma holds. Otherwise, let row l be the highest row that contains a red circle with number j. By assumption, the red circles in rows $1, \ldots, l-1$ all have numbers less than j, indicating that

$$l - 1 \leqslant e_1 + e_2 + \cdots + e_{j-1} \leqslant 1 + 2 + \cdots + 2^{j-2} = 2^{j-1} - 1.$$

This gives $l \leqslant 2^{j-1}$. By Lemma 2, $e_j \leqslant l \leqslant 2^{j-1}$.

Now we see that

$$e_1 + e_2 + \cdots + e_N \leqslant 1 + \cdots + 2^{N-1} = 2^N - 1 < n.$$

So, there must be a red circle with number at least $N + 1$, and this corresponds to a ninja path through at least $N + 1$ red circles.

Proof 3 (Edited from Shi Jiahao's solution) Define v_i as the maximum number of red circles visited by a ninja path that finishes at the ith circle of the bottom row. We claim

$$\sum_{i=1}^{n} \frac{1}{2^{v_i}} + \frac{1}{2^{\max\{v_1, v_2, \ldots, v_n\}}} \leqslant 1. \qquad (*)$$

If $(*)$ is true, then we have

$$1 \geqslant \sum_{i=1}^{n} \frac{1}{2^{\max\{v_1, v_2, \ldots, v_n\}}} + \frac{1}{2^{\max\{v_1, v_2, \ldots, v_n\}}} > \frac{n}{2^{\max\{v_1, v_2, \ldots, v_n\}}},$$

and $\max\{v_1, v_2, \ldots, v_n\} \geqslant \lfloor \log_2 n \rfloor + 1$, which is the problem statement.

We prove $(*)$ by induction. When $n = 1$, $v_1 = 1$ and $(*)$ holds. Suppose $(*)$ is true for $n - 1$. Define $w_j (1 \leqslant j \leqslant n - 1)$ as the maximum number of red circles visited by a (generalized) ninja path that finishes at the jth circle of row $n - 1$. Assume the tth circle in row n is red. Then for $i \neq t$, $v_i = \max\{w_{i-1}, w_i\}$; $v_t = 1 + \max\{w_{t-1}, w_t\}$ (define $w_0 = w_n = -\infty$). Under these conditions, we claim

$$\sum_{i=1}^{n} \frac{1}{2^{v_i}} + \frac{1}{2^{\max\{v_1, v_2, \ldots, v_n\}}} \leqslant \sum_{i=1}^{n} \frac{1}{2^{w_i}} + \frac{1}{2^{\max\{w_1, w_2, \ldots, w_{n-1}\}}}. \qquad (**)$$

Once proved, it will complete the induction of $(*)$.

Assume $w_{t-1} \leqslant w_t$ (otherwise flip $v_1, v_2, \ldots, v_n$ and $w_1, w_2, \ldots, w_n$ so that t becomes $n+1-t$). If $\max\{w_1, w_2, \ldots, w_{n-1}\} = w_t$, as $\forall i < t, v_i \geqslant w_i$ and $\forall i > t, v_i \geqslant w_{i-1}$, and $v_t = w_t + 1$, we have $\max\{v_1, v_2, \ldots, v_n\} \geqslant 1 + w_t$, and

$$\sum_{i=1}^{n} \frac{1}{2^{v_i}} + \frac{1}{2^{\max\{v_1, v_2, \ldots, v_n\}}}$$

$$\leqslant \left(\sum_{i=1}^{t-1} \frac{1}{2^{w_i}} \right) + \frac{1}{2^{w_t+1}} + \left(\sum_{i=t+1}^{n} \frac{1}{2^{w_{i-1}}} \right) + \frac{1}{2^{w_t+1}}$$

$$= \sum_{i=1}^{n-1} \frac{1}{2^{w_i}} + \frac{1}{2^{w_t}},$$

which verifies $(**)$.

If $\max\{w_1, w_2, \ldots, w_{n-1}\} > w_t$ (and $w_t \geqslant w_{t-1}$), from the relation between v_i and w_i we know $\max\{v_1, v_2, \ldots, v_n\} = \max\{w_1, w_2, \ldots, w_{n-1}\}$. There are two situations.

(1) If $\max\{w_1, w_2, \ldots, w_{t-1}\} > w_t$. Suppose that $w_t \geqslant w_{t-1} \geqslant \cdots \geqslant w_k$ and $w_k < w_{k-1}$ $(2 \leqslant k \leqslant t-1)$. Then for any $i < t$ and $i \neq k$, we have $v_i \geqslant w_i$ and $v_k \geqslant w_{k-1} \geqslant w_k + 1$. On the other hand, for any $i > t$, we have $v_i \geqslant w_{i-1}$ and $v_t = w_t + 1$. Hence,

$$\sum_{i=1}^{n} \frac{1}{2^{v_i}} + \frac{1}{2^{\max\{v_1, v_2, \ldots, v_n\}}}$$

$$\leqslant \left(\sum_{i=1}^{t-1} \frac{1}{2^{w_i}} - \frac{1}{2^{w_k}} + \frac{1}{2^{w_k+1}} \right) + \frac{1}{2^{w_t+1}} + \left(\sum_{i=t+1}^{n} \frac{1}{2^{w_{i-1}}} \right)$$

$$+ \frac{1}{2^{\max\{w_1, w_2, \ldots, w_{n-1}\}}}$$

$$= \sum_{i=1}^{n-1} \frac{1}{2^{w_i}} + \frac{1}{2^{\max\{w_1, w_2, \ldots, w_{n-1}\}}} + \frac{1}{2^{w_t+1}} - \frac{1}{2^{w_k+1}}.$$

Since $w_t \geqslant w_{t-1} \geqslant \cdots \geqslant w_k$, $(**)$ holds.

(2) If $\max\{w_{t+1}, w_{t+2}, \ldots, w_{n-1}\} > w_t$. Suppose that $w_t \geqslant w_{t+1} \geqslant \cdots \geqslant w_k$ and $w_k < w_{k+1}$ $(t \leqslant k \leqslant n-2)$. Then for any $i < t$, $v_i \geqslant w_i$ while for any $i > t, i \neq k+1$, $v_i \geqslant w_{i-1}$, and we have $v_{k+1} \geqslant w_{k+1} \geqslant w_k + 1, v_t = w_t + 1$. Therefore,

$$\sum_{i=1}^{n} \frac{1}{2^{v_i}} + \frac{1}{2^{\max\{v_1, v_2, \ldots, v_n\}}}$$

$$\leqslant \left(\sum_{i=1}^{t-1} \frac{1}{2^{w_i}} \right) + \frac{1}{2^{w_t+1}} + \left(\sum_{i=t+1}^{n} \frac{1}{2^{w_{i-1}}} - \frac{1}{2^{w_k}} + \frac{1}{2^{w_k+1}} \right)$$

$$+ \frac{1}{2^{\max\{w_1, w_2, \ldots, w_{n-1}\}}}$$

$$= \sum_{i=1}^{n-1} \frac{1}{2^{w_i}} + \frac{1}{2^{\max\{w_1, w_2, \ldots, w_{n-1}\}}} + \frac{1}{2^{w_t+1}} - \frac{1}{2^{w_k+1}}.$$

Since $w_t \geqslant w_{t+1} \geqslant \cdots \geqslant w_k$, $(**)$ holds.

This completes the proof.

Proof 4 (Edited from Sun Qi'ao's solution) We show that a ninja path traverses at least $\lfloor \log_2 n \rfloor + 1$ red circles. If we can show a ninja path traverses t red circles in a triangle of n rows, then it is also true in a triangle of $n + 1$ rows. So, it suffices to show, in a Japanese triangle of

$n = 2^k$ ($k \in \mathbf{Z}_+$) rows, there is a ninja path through at least $k + 1$ red circles.

Now, there is a red circle in each row for a total of 2^k red circles. Consider a relation "$\to$" between red circles: $A \to B$ if and only if a ninja path can go through A then B; in particular, $A \to A$ is always true. It is not difficult to check $\to$ is a partial order.

A ninja path traverses red circles $A_1, A_2, \ldots, A_l$ in order; it corresponds to a chain $A_1 \to A_2 \to \cdots A_l$. We prove the length of the longest chain is at least $k + 1$; by Mirsky's theorem, equivalently, we prove the minimum number of antichains in a partition is at least $k + 1$. Instead, suppose that the 2^k red circles can be partitioned into at most k antichains.

Lemma *If the highest red circle in an antichain is in row i, then the length of this antichain is at most i.*

Proof of Lemma Let (x, y) be the xth circle (counted from left to right) in row $x + y$, in which $x \in \mathbf{Z}_+, y \in \mathbf{N}$. Then $(x_1, y_1) \to (x_2, y_2)$ if and only if $x_2 \geqslant x_1$ and $y_2 \geqslant y_1$, that is, (x_2, y_2) lies in the equilateral triangle whose top circle is (x_1, y_1).

Suppose the highest red circle is $(t, s), t + s = i$. Any other red circle (x, y) in this antichain must have either $x < t$ or $y < s$. Note that if the first coordinates of A and B are equal, then we must have either $A \to B$ or $B \to A$; it is similar for the second coordinates. This means the antichain has at most $t - 1$ red circles whose first coordinates are $1, 2, \ldots, t - 1$, respectively, and another s red circles whose second coordinates are $0, 1, \ldots, s - 1$, respectively; the length of the antichain is at most $1 + (t - 1) + s = i$. The lemma is proved.

For the original problem, assume the highest red circles in these $t \leq k$ antichains $A_1, A_2, \ldots, A_t$ are located in rows $a_1 < a_2 < \cdots < a_t$, respectively. According to the lemma, we have

$$a_1 = 1 \Rightarrow |A_1| = 1;$$

$$\Rightarrow a_2 = 2 \Rightarrow |A_2| \leqslant 2;$$

$$\Rightarrow a_3 \leqslant 4 \Rightarrow |A_3| \leqslant 4;$$

$$\cdots \quad \cdots$$

$$\Rightarrow a_t \leqslant 2^{t-1} \Rightarrow |A_t| \leqslant 2^{t-1}.$$

This leads to $|A_1| + \cdots + |A_t| \leqslant 2^t - 1 < 2^k$, a contradiction. (Note that $A_1, \ldots, A_t$ are antichains of a partition.)

In all, the answer is $\lfloor \log_2 n \rfloor + 1$.

Remark   We briefly introduce Dilworth's theorem and Mirsky's theorem in finite partially ordered sets. Given a partially ordered set $(P, \prec)$, a chain consists of elements any two of which are comparable; an antichain consists of elements any two of which are not comparable. For example, $P = \{1, 2, \ldots, 8\}$, $a \prec b$ if and only if $a|b$: $\{1, 3, 6\}$ is a chain; $\{1\}$ and $\{2, 5, 7\}$ are antichains.

Dilworth's theorem   In $(P, \prec)$, the length of the longest antichain equals the minimum number of chains in a partition of P (the chains are disjoint and their union is P). For example, $P = \{1, 2, \ldots, 8\}$, the length of the longest antichain, say $\{2, 3, 5, 7\}$ or $\{4, 5, 6, 7\}$, is 4; the minimum number of chains in a partition of P is also 4: $P = \{1, 2, 4, 8\} \cup \{3, 6\} \cup \{5\} \cup \{7\}$.

Mirsky's theorem   In $(P, \prec)$, the length of the longest chain equals the minimum number of antichains in a partition of P. For example, $P = \{1, 2, \ldots, 8\}$, the length of the longest chain $\{1, 2, 4, 8\}$ is 4; the minimum number of antichains in a partition is also 4: $P = \{1\} \cup \{2, 3\} \cup \{4, 5, 6, 7\} \cup \{8\}$.

The two theorems are dual to each other.

Proof of Dilworth's theorem   First, in any chain partition $C_1, \ldots, C_k$ of P, an antichain A contains at most one element from each C_i. Hence, $|A| \leqslant k$.

Next, we apply induction to prove the existence of an antichain whose length equals the number of chains in a partition of P. The statement is true for empty and single element sets. For P with $|P| \geqslant 2$, consider $P' = P \backslash \{a\}$. By the induction hypothesis, there exist $P' = C_1 \cup C_2 \cup \cdots \cup C_k$ and antichain A_0 with $|A_0| = k$. Evidently, $A_0 \cap C_i \neq \varnothing$ for $i = 1, 2, \ldots, k$. Let

$$x_i = \max\{x \in C_i \cap A : A \text{ is an antichain}, |A| = k\},$$

$$1 \leqslant i \leqslant k, \quad \text{and} \quad A_{\max} = \{x_1, x_2, \ldots, x_k\}.$$

Here, x_i is well defined because C_i is totally ordered. We claim $A_{\max}$ is an antichain. Suppose for $1 \leqslant i \leqslant k$, $x_i \in A_i$, A_i is an antichain, $|A_i| = k$. For indices $i \neq j$, $A_i \cap C_j \neq \varnothing$. Take $y \in A_i \cap C_j$. By definition of x_j, we have $y \leqslant x_j$, yet $x_i \not\geqslant y$, and thus $x_i \not\geqslant x_j$. Interchanging i, j, we get $x_j \not\geqslant x_i$. This shows $A_{\max}$ is an antichain.

Back to P. If $a \geqslant x_i$ for some $i \in \{1, 2, \ldots, k\}$, let K be the chain $\{a\} \cup \{z \in C_i | z \leqslant x_i\}$. By definition of x_i, the set $P \backslash K$ does not contain any antichain of length k. By the induction hypothesis, $P \backslash K$ can be covered by $k - 1$ disjoint chains, and thus P can be covered by k disjoint chains, completing the inductive proof for this situation. Otherwise, $a \not\geq x_i$ for every $i \in \{1, 2, \ldots, k\}$. Then the set $A_{\max} \cup \{a\}$ is an antichain of length $k + 1$ in P, and P can be covered by $k + 1$ chains $\{a\}, C_1, C_2, \ldots, C_k$. This completes the proof.

Proof of Mirsky's theorem First, it is obvious that the length k of the longest chain cannot exceed the minimum number of antichains in a partition of P. The following constructive proof for necessity is simpler than that of Dilworth's theorem. For each element $x \in P$, let

$$N(x) = \max |C| : C \text{ is a chain and } x \in C \text{ is the largest element of } C.$$

It is straightforward to check that each set $N^{-1}(i)$, consisting of elements with the same N value as i, is an antichain. These k antichains $N^{-1}(1), N^{-1}(2), \ldots, N^{-1}(k)$ form a partition of P.

Problem 6 Let ABC be an equilateral triangle. Let A_1, B_1, C_1 be interior points of ABC such that $BA_1 = A_1C, CB_1 = B_1A, AC_1 = C_1B$, and

$$\angle BA_1C + \angle CB_1A + \angle AC_1B = 480°.$$

Let BC_1 and CB_1 meet at point A_2; CA_1 and AC_1 meet at point B_2; AB_1 and BA_1 meet at point C_2. Prove that if triangle $A_1B_1C_1$ is scalene, then the three circumcircles of triangles AA_1A_2, BB_1B_2, and CC_1C_2 all pass through two common points.

Note: A scalene triangle is one where no two sides have equal length.

(Contributed by United States of America)

We give six solutions as below.

Solution 1 Denote the circumcircles of $\triangle AA_1A_2, \triangle BB_1B_2, \triangle CC_1C_2$ by $\delta_A, \delta_B, \delta_C$, respectively. We need to find two different points that have the same power with respect to $\delta_A, \delta_B, \delta_C$.

Lemma 1. A_1 *is the circumcenter of* $\triangle A_2BC$; B_1 *is the circumcenter of* $\triangle B_2CA$; C_1 *is the circumcenter of* $\triangle C_2AB$.

Proof of Lemma 1 We only prove the first part; the other two are analogous. As illustrated in Fig. 10.7, A_1 is inside $\triangle BA_2C$ and on the

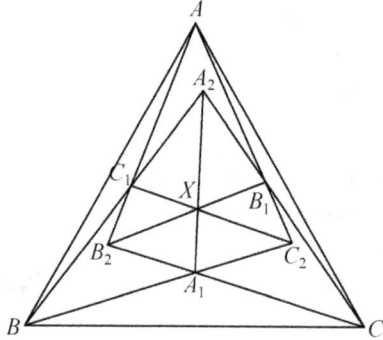

Fig. 10.7

perpendicular bisector of BC; it only requires $\angle BA_1C = 2\angle BA_2C$. We have

$$\angle BA_2C = \angle A_2BA + \angle BAC + \angle ACA_2$$

$$= \frac{1}{2}(180° - \angle AC_1B) + 60° + \frac{1}{2}(180° - \angle CB_1A)$$

$$= 240° - \frac{1}{2}(480° - \angle BA_1C) = \frac{1}{2}\angle BA_1C.$$

Lemma 1 gives

$$\angle B_1B_2C_1 = \angle B_1B_2A = \angle B_2AB_1 = \angle C_1AC_2 = \angle AC_2C_1 = \angle B_1C_2C_1.$$

It follows that B_1, C_1, B_2, C_2 are concyclic, and likewise C_1, A_1, C_2, A_2 and A_1, B_1, A_2, B_2. Note that the hexagon $A_1B_2C_1A_2B_1C_2$ is not cyclic because $\angle C_2A_1B_2 + \angle B_2C_1A_2 + \angle A_2B_1C_2 = 480° \neq 360°$. By the radical axis theorem, the lines A_1A_2, B_1B_2 and C_1C_2 intersect at one point say X, and the powers of X with respect to $\delta_A, \delta_B, \delta_C$ are equal.

Let $A_3 \neq A_2$ be the other intersection point of the circumcircle of $\triangle A_2BC$ and δ_A; define B_3 and C_3 in a similar way.

Lemma 2   *The four points B, C, B_3, C_3 are concyclic.*

Proof of Lemma 2   As shown in Fig. 10.8, we use directed angles:

$$\angle BC_3C = \angle BC_3C_2 + \angle C_2C_3C$$

$$= \angle BAC_2 + \angle C_2C_1C$$

$$= 90° + \angle(C_1C, AC_2) + \angle C_2C_1C \text{ (since } CC_1 \perp AB)$$

$$= 90° + \angle C_1C_2B_1.$$

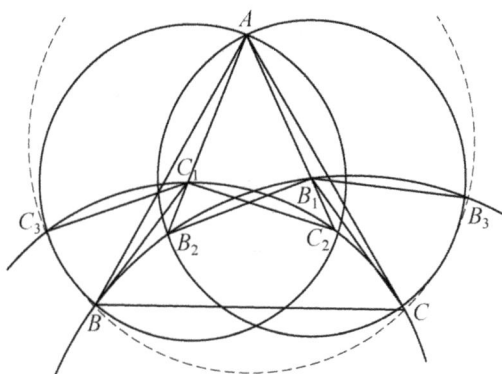

Fig. 10.8

Similarly, $\angle CB_3B = 90° + \angle B_1B_2C_1$. Since B_1, C_1, B_2, C_2 are concyclic, we reach

$$\angle BB_3C = 90° + \angle C_1B_2B_1 = 90° + \angle C_1C_2B_1 = \angle BC_3C,$$

and the lemma is proved.

By similar arguments, CAC_3A_3 and ABA_3B_3 are cyclic quadrilaterals as well. However, $AC_3BA_3CB_3$ is not a cyclic hexagon, or else A, B_2, C, B_3 are concyclic, B_2 lies on the circumcircle of ABC, but B_2 is inside $\triangle ABC$, so this is impossible. Apply the radical axis theorem to these three circles to find AA_3, BB_3, CC_3 intersect at one point say Y, and the powers of Y to $\delta_A, \delta_B, \delta_C$ are equal.

We need to discuss the locations of X, Y. Let O be the center of $\triangle ABC$. Since

$$\angle BA_1C = 480° - \angle CB_1A - \angle AC_1B > 480° - 180° - 180° = 120°,$$

A_1 lies inside $\triangle BOC$; it is similar for B_1 and C_1. Thus, $\triangle BA_1C, \triangle CB_1A$ and $\triangle AC_1B$ have non-overlapping interior, which implies that $A_1B_2C_1 A_2B_1C_2$ is a convex hexagon, X is on segment A_1A_2, which is inside δ_A.

Since A_1 is the circumcenter of $\triangle A_2BC$, $A_1A_2 = A_1A_3$. Since A, A_2, A_1, A_3 are concyclic, line AA_2 is the reflection of $AA_3 \equiv AY$ about AA_1. As X lies on segment A_1A_2, the only way we can possibly have $X \equiv Y$ is that both A_1 and A_2 lie on the perpendicular bisector of BC, but this means B_1 and C_1 are symmetric about AA_1, $A_1B_1 = A_1C_1$, which violates the given scalene condition.

In summary, we have obtained two different points X, Y that have the same power to $\delta_A, \delta_B, \delta_C$. Hence, these circles have a common radical axis.

As X is inside δ_A (and inside δ_B, δ_C as well), this radical axis intersects the circle at two points. We conclude that $\delta_A, \delta_B, \delta_C$, or the circumcircles of $\triangle AA_1A_2, \triangle BB_1B_2, \triangle CC_1C_2$, have two common points of intersection.

Solution 2 As in Solution 1, we have three different circumcircles $\omega_A = \odot B_1B_2C_1C_2, \omega_B = \odot C_1C_2A_1A_2$, and $\omega_C = \odot A_1A_2B_1B_2$. Define

$$\Gamma_A = \frac{Pow_{\omega_B}(A)}{Pow_{\omega_C}(A)}, \quad \Gamma_B = \frac{Pow_{\omega_C}(B)}{Pow_{\omega_A}(B)}, \quad \Gamma_C = \frac{Pow_{\omega_A}(C)}{Pow_{\omega_B}(C)}.$$

By the coaxial circles lemma, $\delta_A = \odot AA_1A_2$ is the trace of point Z that satisfies $Pow_{\omega_B}(Z) = \Gamma_A \cdot Pow_{\omega_C}(Z)$.

Lemma 3 $\Gamma_A\Gamma_B\Gamma_C = 1$.

Proof of Lemma 3 Let $\alpha = \angle A_1BC, \beta = \angle B_1CA, \gamma = \angle C_1AB$. The problem condition implies $\alpha + \beta + \gamma = 30°$. Let $A_B = AA_1 \cap \omega_C$ and $A_C = AA_1 \cap \omega_B$. Then

$$\angle A_1AB_1 = 30° - \beta, \quad \angle A_1A_BB_1 = \angle A_1A_2B_1 = \angle B_1CA_1 = 30° + \gamma,$$

and thus $\angle AB_1A_B = \gamma + \beta = 30° - \alpha$.

The sine rule gives $\frac{AA_B}{AB_1} = \frac{\sin(30° - \alpha)}{\sin(30° + \gamma)}$. Express $\frac{AA_C}{AC_1}$ in a similar way and we obtain

$$\Gamma_A = \frac{AA_C \cdot AA_1}{AA_B \cdot AA_1} = \frac{AC_1\sin(30° + \gamma)}{AB_1\sin(30° + \beta)}.$$

Together with similar expressions for Γ_B, Γ_C and conditions $BA_1 = CA_1$, etc., we reach $\Gamma_A\Gamma_B\Gamma_C = 1$.

Now, for any point Y that lies on $\delta_A \cap \delta_B$,

$$Pow_{\omega_B}(Y) \overset{y \in \delta_A}{=} \Gamma_A Pow_{\omega_C}(Y) \overset{y \in \delta_B}{=} \Gamma_A\Gamma_B Pow_{\omega_A}(Y)$$

$$\Rightarrow Pow_{\omega_A}(Y) = \Gamma_C Pow_{\omega_B}(Y)$$

$$\Rightarrow Y \in \delta_C.$$

So, $Y \in (\delta_A \cap \delta_B \cap \delta_C)$. Finally, adopt the same argument as in Solution 1 to clarify that δ_A and δ_B have two different intersection points, completing the proof.

Solution 3 (Edited from Liang Xingjian's solution) As in Solution 1, we see that A_1, B_1, C_1 are the circumcenters of triangles A_2BC, B_2AC, C_2AB, respectively; AA_1, BB_1, CC_1 are concurrent, say they intersect at X; the powers of X to $\odot AA_1A_2, \odot BB_1B_2, \odot CC_1C_2$ are equal. Since X lies on the radical axis of $\odot BB_1B_2$ and $\odot CC_1C_2$, X is inside $\odot BB_1B_2$, thus $\odot BB_1B_2$ and $\odot CC_1C_2$ intersect say at P and Q.

Since PQ is the radical axis of $\odot BB_1B_2$ and $\odot CC_1C_2$, P, X, Q are collinear, and

$$PX \cdot XQ = B_1X \cdot B_2X = A_1X \cdot A_2X.$$

So, A_1, P, A_2, Q are concyclic. To show $\odot AA_1A_2$ also passes through P and Q, it only requires A, A_1, P, Q concyclic.

We introduce an inversion transformation centered at A with arbitrary power. After the transformation, A_1 lies on the perpendicular bisector of BC, A_1 is a point of intersection of $\odot ABC_2$ and $\odot ACB_2$; moreover, C_1, B_1 are the symmetric points of A about BC_2, CB_2, respectively. It is equivalent to proving that A_1, P, Q are collinear, that is, A_1 lies on the radical axis of $\odot BB_1B_2$ and $\odot CC_1C_2$.

As illustrated in Fig. 10.9, let $\odot BB_1B_2$ and $\odot CC_1C_2$ intersect lines BA_1 and CA_1 again at points U and V, respectively. As $BA_1 = CA_1$, we only need to prove U, V are symmetric about AA_1. Note that C is the circumcenter of $\triangle ABB_1$, and thus

$$\angle B_2UB = \angle B_2B_1B = \angle B_2B_1A - \angle BB_1A = \angle C_1AB_1 - 30°.$$

Similarly, $\angle C_2VC = \angle C_1AB_1 - 30°$, so $\angle B_2UB = \angle C_2VC$. It suffices to prove the symmetric point C_2' of C_2 about AA_1 lies on B_2U. Since

$$\angle ABU = 150° - \angle BA_1A = 150° - \angle BC_2A = 60° + \angle C_1AB_1 = 90° + \angle B_2UB,$$

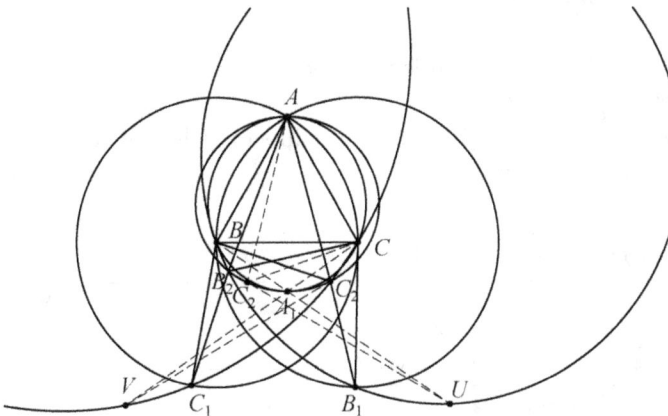

Fig. 10.9

$AB \perp B_2 U$. Since

$$\angle AC_2'C = \angle AC_2 B = 90° - \angle C_1 AB_1 = \angle AB_2 C,$$

A, C, C_2', B_2 are concyclic. Therefore,

$$\angle AB_2 C_2' = 180° - \angle ACC_2' = 180° - \angle ABC_2 = 90° + \angle BAB_2.$$

This implies $B_2 C_2' \perp AB$ and C_2' lies on $B_2 U$, completing the proof.

Solution 4 (Edited from Jiang Zhicheng and Wang Chunji's solutions)

We establish the complex plane with unit circle $\odot ABC$; use a lowercase letter to represent the complex number corresponding to the point of the uppercase letter. Let

$$\angle A_1 CB = \angle A_1 BC = \theta_1, \quad \angle B_1 CA = \angle B_1 AC = \theta_2,$$
$$\angle C_1 AB = \angle C_1 BA = \theta_3; \quad \alpha = e^{2i\theta_1}, \quad \beta = e^{2i\theta_2}, \quad \gamma = e^{2i\theta_3}.$$

From the problem, we have $\theta_1 + \theta_2 + \theta_3 = 30°$, $\alpha\beta\gamma = -\omega^2$ where $\omega = e^{\frac{2\pi i}{3}}$, and $b = a\omega, c = a\omega^2$.

Note that the second intersection points of BA_1, CA_1 and $\odot ABC$ are αc and $\frac{b}{\alpha}$, respectively. Hence,

$$a_1 = \frac{b\frac{c}{\alpha}(b + c\alpha) - c(b\alpha)\left(c + \frac{b}{\alpha}\right)}{b\frac{c}{\alpha} - c(b\alpha)} = \frac{\omega + \omega^2\alpha - \omega\alpha - \omega^2\alpha^2}{1 - \alpha^2}a = \frac{\omega^2\alpha + \omega}{1 + \alpha}a.$$

(The formula for the intersection point of two chords xy and zw of the unit circle: $\frac{xy(z+w) - zw(x+y)}{xy - zw}$.)

Moreover, the second intersection points of BA_2, CA_2 and $\odot ABC$ are $\frac{\alpha}{\gamma}$ and βa, respectively. Hence,

$$a_2 = \frac{c(\beta a)\left(\frac{a}{\gamma} + \beta\right) - b\left(\frac{a}{\gamma}\right)(c + \beta a)}{c(\beta a) - b\left(\frac{a}{\gamma}\right)} = \frac{\omega^2\beta\left(\frac{1}{\gamma} + \beta\right) - \omega\left(\frac{1}{\gamma}\right)(\omega^2 + \beta)}{\omega^2\beta - \omega\left(\frac{1}{\gamma}\right)}a$$

$$= \frac{\beta\gamma + \omega^2\beta - \omega\beta - 1}{\omega^2\beta\gamma - \omega}a = \frac{\alpha + (\omega - \omega^2)\alpha\beta + \omega^2}{\omega(1 + \alpha)}a. \tag{*}$$

Here, $\beta\gamma = -\frac{\omega^2}{\alpha}$.

Next, we try to find the equation for $\odot AA_1A_2$:

$$\frac{(Z-a)(a_1-a_2)}{(Z-a_2)(a_1-a)} \in \mathbf{R}$$

$$\Leftrightarrow \frac{(Z-a)(a_1-a_2)}{(Z-a_2)a} \in \mathbf{R}$$

$$\Leftrightarrow \frac{(Z-a)(\alpha+\omega^2-\alpha-(\omega-\omega^2)\alpha\beta-\omega^2)}{(Z-a_2)(\omega(1+\alpha))} \in \mathbf{R}$$

$$\Leftrightarrow \frac{(Z-a)(\alpha\beta)}{(Z-a_2)(\omega(1+\alpha))} \in i\mathbf{R} \Leftrightarrow \frac{(Z-a)\omega}{(Z-a_2)\gamma(1+\alpha)} \in i\mathbf{R}$$

$$\Leftrightarrow (Z-a)\omega(\overline{Z}-\overline{a}_2)\overline{\gamma}(1+\overline{\alpha})+(Z-a_2)\gamma(1+\alpha)(\overline{Z}-\overline{a})\overline{\omega}=0$$

$$\Leftrightarrow (Z-a)(\overline{Z}-\overline{a}_2)+\omega\gamma^2\alpha(\overline{Z}-\overline{a})(Z-a_2)=0$$

$$\Leftrightarrow \beta(Z-a)(\overline{Z}-\overline{a}_2)-\gamma(\overline{Z}-\overline{a})(Z-a_2)=0$$

$$\Leftrightarrow (\beta-\gamma)Z\overline{Z}-(\overline{a}_2\beta-\gamma\overline{a})Z-(\beta a-\gamma a_2)\overline{Z}+(\beta a\overline{a}_2-\gamma a_2\overline{a})=0. \tag{$**$}$$

To eliminate a_2 in the coefficients, use $(*)$ and its conjugate form $\overline{a}_2 = \frac{\beta+(\omega^2-\omega)+\omega^2\alpha\beta}{\omega^2(1+\alpha)\beta}\overline{a}$:

$$\overline{a}_2\beta-\gamma\overline{a} = \frac{\overline{a}}{(1+\alpha)}[\omega\beta+(1-\omega^2)+\omega^2\alpha\beta-\gamma-\gamma\alpha]$$

$$= \frac{\beta\overline{c}-\gamma\overline{a}+\overline{b}\alpha\beta-\overline{a}\gamma\alpha+(1-\omega^2)\overline{a}}{1+\alpha};$$

$$\beta a-\gamma a_2 = \frac{\beta a+\alpha\beta a-\omega^2 a\alpha\gamma+(1-\omega)a\alpha\beta\gamma-\omega\gamma a}{1+\alpha}$$

$$= \frac{\beta a-\gamma b+\alpha\beta a-\gamma\alpha c+(1-\omega^2)a}{1+\alpha};$$

$$\beta a\overline{a}_2-\gamma a_2\overline{a} = \frac{\beta+\omega\alpha\beta+(\omega^2-\omega)}{\omega^2(1+\alpha)}-\frac{\gamma[\alpha+(\omega-\omega^2)\alpha\beta+\omega^2]}{\omega(1+\alpha)}$$

$$= \frac{(\beta-\gamma)(1+\omega\alpha)}{\omega(1+\alpha)}.$$

Here, $\alpha\beta\gamma=-\omega^2$ is used. Substitute them into $(**)$ to get the equation for $\odot AA_1A_2$:

$$(1+\alpha)(\beta-\gamma)Z\overline{Z}-[\beta\overline{c}-\gamma\overline{a}+\overline{b}\alpha\beta-\overline{a}\gamma\alpha+(1-\omega^2)\overline{a}]Z$$

$$-[\beta a-\gamma b+\alpha\beta a-\gamma\alpha c+(1-\omega^2)a]\overline{Z}+\omega^2(\beta-\gamma)(1+\omega\alpha)=0.$$

Now we turn to the equations for $\odot BB_1B_2$ and $\odot CC_1C_2$: the coefficients of $Z\overline{Z}, Z, \overline{Z}$ and 1 are permutations of α, β, γ and a, b, c of the above. The key is to note that every cyclic sum of the coefficients adds to zero, which implies that the sum of the three equations is $0 = 0$. If Z satisfies two equations, then it must satisfy the third equation. This means the intersection of two circles also lies on the third circle.

It only requires to show there are two intersection points. As the centre O is on segments AA_1, BB_1, CC_1, the three circles have common interior points. We show that they are not mutually inclusive. As B_2 is inside quadrilateral BC_1B_1C, if C_1 and C are both on or inside $\odot BB_1B_2$, then B_2 is inside $\odot BB_1B_2$, a contradiction. Thus, $\odot BB_1B_2$ does not contain or touch $\odot CC_1C_2$. The same is true for any other pair of the circles. Therefore, the circles are not mutually inclusive; they have two intersection points, and the conclusion follows.

Remark Instead of calculating a_1, one can also derive the equation $(**)$ for $\odot AA_1A_2$ by calculating the directed angle $\angle AA_1A_2 = \beta - \gamma$.

Solution 5 (Edited from Shi Haojia's solution) Let O be the center of $\triangle ABC$. As in Solution 1, A_1, B_1, C_1 are the centers of $\odot A_2BC, \odot B_2CA, \odot C_2AB$, respectively.

Define $\angle AC_1B = 2\alpha, \angle BA_1C = 2\beta, \angle AB_1C = 2\gamma$. Then $\alpha + \beta + \gamma = \frac{4\pi}{3}$ and $\alpha, \beta, \gamma \in \left(\frac{\pi}{3}, \frac{\pi}{2}\right)$. Note that $\angle AC_1C > \frac{\pi}{2} > \frac{\angle AB_1C}{2} = \angle AB_2C$, and thus C_1 is on segment AB_2; similarly, A_1 is on segment CB_2. Assume that B_2 is inside $\triangle AOB$ (as $\angle B_2AB = \angle C_1AB < \angle OAB$, B_2 is not inside $\triangle AOC$). We have

$$\angle BB_1B_2 = \frac{\angle CB_1B_2 - \angle AB_1B_2}{2} = \angle CAB_2 - \angle ACB_2$$

$$= \angle B_2CB - \angle B_2AB = \left(\frac{\pi}{2} - \beta\right) - \left(\frac{\pi}{2} - \alpha\right) = \alpha - \beta.$$

Due to symmetry, if any two of α, β, γ are equal, then two sides of $\triangle A_1B_1C_1$ are equal, contradiction. So, α, β, γ are distinct, $\alpha > \beta$, and $\angle BB_1B_2 > 0$. This implies B, B_1, B_2 are not collinear, and similarly A, A_1, A_2 are not collinear, C, C_1, C_2 are not collinear, either.

Claim: $\odot BB_1B_2$ and $\odot CC_1C_2$ have two different intersection points.

Let $S \neq B_2$ be the second intersection point of $\odot BB_1B_2$ and B_2C. Then $\angle BSB_2 = \angle BB_1B_2 = \alpha - \beta < \frac{\pi}{2} - \beta = \angle A_1CB = \angle B_2CB$, implying that C is on segment B_2S and $C \neq S, B_2$. Since S, B_2 are on $\odot BB_1B_2$, C must be inside $\odot BB_1B_2$. Since $B_2 \in \triangle AOB \subseteq \triangle AB_1B$ and C_1 is on segment AB_2, $\angle BC_1B_1 = \angle BB_2B_1 - \angle B_2BC_1 - \angle B_2B_1C_1 < \angle BB_2B_1$. This indicates that C_1 is outside $\odot BB_1B_2$, and thereby $\odot CC_1C_2$ and $\odot BB_1B_2$ have two different points of intersection, say L_1 and L_2.

For point X, denote the power of X with respect to $\odot AA_1A_2$ by $\rho_A(X)$, and likewise $\rho_B(X)$ and $\rho_C(X)$. In any rectangular coordinate system, the pairwise differences of $\rho_A(X), \rho_B(X)$ and $\rho_C(X)$ are linear functions of the two coordinates (as the $x^2 + y^2$ terms cancel out). If $\overrightarrow{OP} = \lambda\overrightarrow{OA} + \mu\overrightarrow{OB} + \tau\overrightarrow{OC}$ with $\lambda + \mu + \tau = 1$, then

$$\rho_A(P) - \rho_B(P) = \lambda(\rho_A(A) - \rho_B(A)) + \mu(\rho_A(B) - \rho_B(B))$$
$$+ \tau(\rho_A(C) - \rho_B(C));$$

$$\rho_B(P) - \rho_C(P) = \lambda(\rho_B(A) - \rho_C(A)) + \mu(\rho_B(B) - \rho_C(B))$$
$$+ \tau(\rho_B(C) - \rho_C(C)).$$

We will prove

$$\frac{\rho_A(A) - \rho_B(A)}{\rho_B(A) - \rho_C(A)} = \frac{\rho_A(B) - \rho_B(B)}{\rho_B(B) - \rho_C(B)} = \frac{\rho_A(C) - \rho_B(C)}{\rho_B(C) - \rho_C(C)},$$

in which the denominators are all nonzero. It will follow that $\rho_A(P) - \rho_B(P)$ is a constant multiplied by $\rho_B(P) - \rho_C(P)$, and $\rho_B(P) = \rho_C(P)$ will imply $\rho_A(P) = \rho_B(P)$. Taking $P = L_1, L_2$, we have $\rho_B(P) = \rho_C(P) = 0$, then $\rho_A(P) = 0$, L_1, L_2 are on $\odot AA_1A_2$, and this will complete the proof.

Obviously, $\rho_A(A) = \rho_B(B) = \rho_C(C) = 0$. Now compute $\rho_B(A), \rho_A(B)$, $\rho_B(C), \rho_C(B), \rho_C(A), \rho_A(C)$. (It is not required that B_2 lies inside $\triangle OAB$; use directed angles for all angles involved.) We have

$$\rho_B(C) = -CS \cdot CB_2 = -BC \cdot \frac{\sin \angle SBC}{\sin \angle BSC} \cdot AC \cdot \frac{\sin \angle CAB_2}{\sin \angle AB_2C}$$

$$= -AB^2 \cdot \frac{\sin\left(\frac{\pi}{2} - \beta - \angle BSB_2\right)}{\sin \angle BSB_2} \cdot \frac{\sin\left(\frac{\pi}{3} - \left(\frac{\pi}{2} - \alpha\right)\right)}{\sin \frac{\angle AB_1C}{2}}$$

$$= -AB^2 \cdot \frac{\cos\alpha \cdot \sin\left(\alpha - \frac{\pi}{6}\right)}{\sin\left(\alpha - \beta\right) \cdot \sin\gamma},$$

(here $\angle BSB_2 = \alpha - \beta$ was shown) and similar expressions for the five powers. Once

$$\frac{\rho_A(A) - \rho_B(A)}{\rho_B(A) - \rho_C(A)} = \frac{\rho_A(B) - \rho_B(B)}{\rho_B(B) - \rho_C(B)} \qquad \text{①}$$

is verified, by interchanging B and C,

$$\frac{\rho_A(A) - \rho_C(A)}{\rho_C(A) - \rho_B(A)} = \frac{\rho_A(C) - \rho_C(C)}{\rho_C(C) - \rho_B(C)},$$

adding 1 to both sides, and multiplying by -1, we obtain

$$\frac{\rho_A(A) - \rho_B(A)}{\rho_B(A) - \rho_C(A)} = \frac{\rho_A(C) - \rho_B(C)}{\rho_B(C) - \rho_C(C)};$$

together with ①, the conclusion follows. Indeed, we have

$$\rho_B(A) = -AB^2 \cdot \frac{\cos\beta \cdot \sin\left(\beta - \frac{\pi}{6}\right)}{\sin(\beta - \alpha) \cdot \sin\gamma};$$

$$\rho_C(A) = -AB^2 \cdot \frac{\cos\beta \cdot \sin\left(\beta - \frac{\pi}{6}\right)}{\sin(\beta - \gamma) \cdot \sin\alpha};$$

$$\rho_B(A) - \rho_C(A) = AB^2 \cdot \left(\frac{\cos\beta \cdot \sin\left(\beta - \frac{\pi}{6}\right)}{\sin(\beta - \gamma) \cdot \sin\alpha} - \frac{\cos\beta \cdot \sin\left(\beta - \frac{\pi}{6}\right)}{\sin(\beta - \alpha) \cdot \sin\gamma}\right)$$

$$= AB^2 \cdot \frac{\cos\beta \cdot \sin\left(\beta - \frac{\pi}{6}\right) \cdot \sin\beta \sin(\gamma - \alpha)}{\sin(\beta - \gamma) \cdot \sin\alpha \cdot \sin(\beta - \alpha) \cdot \sin\gamma} \neq 0,$$

where the last equality comes from

$$\sin(\beta - \alpha) \cdot \sin\gamma - \sin(\beta - \gamma) \cdot \sin\alpha$$

$$= \sin\beta \cdot \cos\alpha \cdot \sin\gamma - \sin\beta \cdot \cos\gamma \cdot \sin\alpha$$

$$= \sin\beta \cdot (\cos\alpha \cdot \sin\gamma - \cos\gamma \cdot \sin\alpha) = \sin\beta \sin(\gamma - \alpha).$$

Moreover,

$$\rho_A(B) = -AB^2 \cdot \frac{\cos\gamma \cdot \sin\left(\gamma - \frac{\pi}{6}\right)}{\sin(\gamma - \alpha) \cdot \sin\beta};$$

$$\rho_C(B) = -AB^2 \cdot \frac{\cos\gamma \cdot \sin\left(\gamma - \frac{\pi}{6}\right)}{\sin(\gamma - \beta) \cdot \sin\alpha} \neq 0.$$

Together, they give

$$\frac{\rho_B(A)}{\rho_B(A) - \rho_C(A)} = -\frac{\sin(\beta - \gamma)\sin\alpha}{\sin\beta \sin(\gamma - \alpha)} = \frac{\rho_A(B)}{\rho_C(B)} \quad \Leftrightarrow \qquad ①$$

as desired.

Solution 6 (Edited from Sun Qi'ao's solution) As in Solution 1, we have

(1) A_1, B_1, C_1 are the centers of $\odot A_2 BC, \odot B_2 CA$, and $\odot C_2 AB$, respectively;

(2) $A_1 A_2, B_1 B_2, C_1 C_2$ have a common point of intersection, say P;

(3) The powers of P to $w_a = \odot AA_1 A_2, w_b = \odot BB_1 B_2$, and $w_c = \odot CC_1 C_2$, are equal;

(4) Each pair of w_a, w_b, w_c has two intersection points.

Now we Prove that the ratios of the powers of A and A_1 to w_b, w_c are equal.

Let $\angle B_2 A C_2 = \alpha, \angle C_2 B A_2 = \beta, \angle A_2 C B_2 = \gamma$ (see Fig. 10.10). The problem gives $\alpha + \beta + \gamma = 120°$, $\alpha, \beta, \gamma \in (30°, 60°)$. By manipulation of angles, we find $\angle AC_2 C_1 = \alpha, \angle B_2 C_1 C_2 = 2\alpha, \angle A_1 BC = \alpha - 30°$, and so on.

Suppose AB and w_b intersect at two points B, X_b; AC and w_c intersect at C, X_c; $A_1 B$ and w_b intersect at B, Y_b; $A_1 C$ and w_c intersect at C, Y_c. We have

$$\angle C_1 C Y_c = \angle C_1 C A_1 = 30° - \angle A_1 CB = 60° - \alpha;$$

$$\angle Y_c C_2 A_1 = \angle C_1 C_2 A_1 - \angle C_1 C_2 Y_c = \beta - (60° - \alpha) = 60° - \gamma;$$

$$\angle C X_c C_2 = \angle C Y_c C_2 = \pi - \angle C_2 A_1 Y_c - \angle Y_c C_2 A_1 = \alpha - \beta;$$

$$\angle A C_2 X_c = \angle A X_c C_2 + \angle C_2 AC = \alpha - 30°.$$

On one hand, the power of A to w_c is

$$\overline{AC} \cdot \overline{AX_c} = AC \cdot AC_2 \cdot \frac{\sin(\alpha - 30°)}{\sin(\alpha - \beta)},$$

which is positive when $\alpha - \beta > 0$ and negative otherwise. The power of A to w_b is

$$\overline{AB} \cdot \overline{AX_b} = AB \cdot AB_2 \cdot \frac{\sin(\alpha - 30°)}{\sin(\alpha - \gamma)},$$

leading to the ratio of powers

$$r_A = \frac{\overline{AB} \cdot \overline{AX_b}}{\overline{AC} \cdot \overline{AX_c}} = \frac{AB_2 \cdot \sin(\alpha - \beta)}{AC_2 \cdot \sin(\alpha - \gamma)} = \frac{AB_1}{AC_1} \cdot \frac{\sin(\alpha - \beta)}{\sin(\alpha - \gamma)}$$

$$= = \frac{\cos \angle C_1 AB}{\cos \angle B_1 AC} \cdot \frac{\sin(\alpha - \beta)}{\sin(\alpha - \gamma)}$$

$$= \frac{\sin(120° - \gamma)}{\sin(120° - \beta)} \cdot \frac{\sin(\alpha - \beta)}{\sin(\alpha - \gamma)}.$$

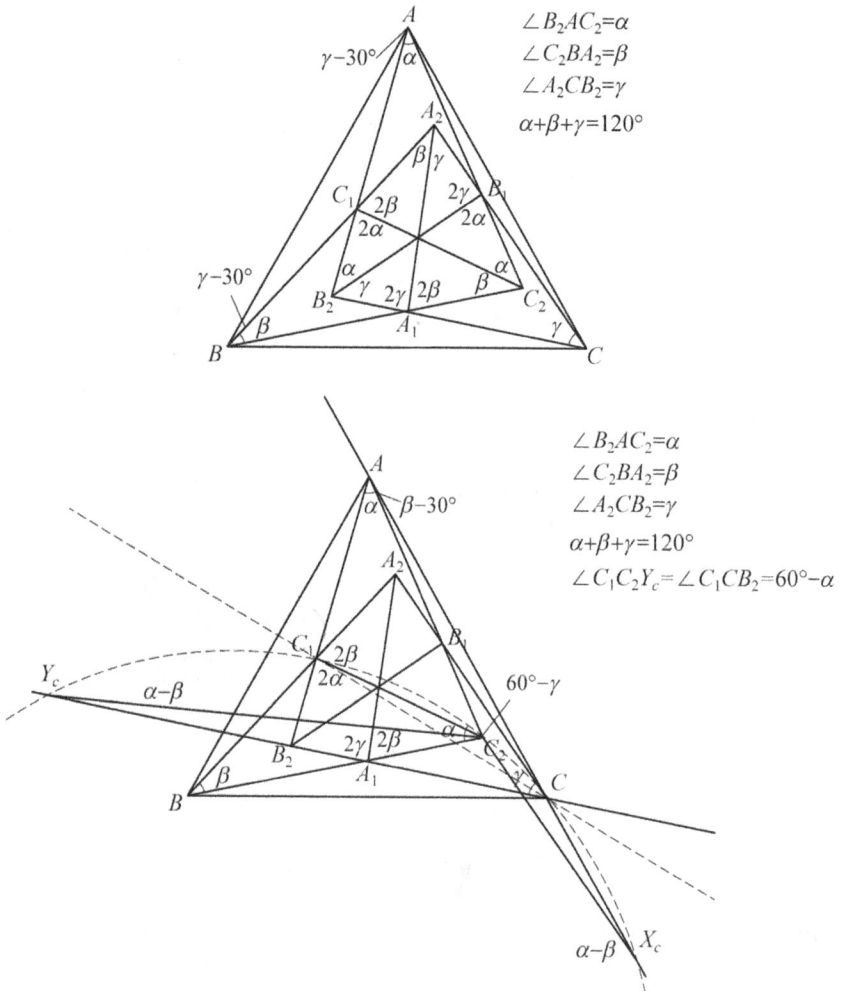

$\angle B_2AC_2=\alpha$
$\angle C_2BA_2=\beta$
$\angle A_2CB_2=\gamma$
$\alpha+\beta+\gamma=120°$

$\angle B_2AC_2=\alpha$
$\angle C_2BA_2=\beta$
$\angle A_2CB_2=\gamma$
$\alpha+\beta+\gamma=120°$
$\angle C_1C_2Y_c=\angle C_1CB_2=60°-\alpha$

Fig. 10.10

On the other hand, the power of A_1 to ω_c is

$$\overline{A_1C} \cdot \overline{A_1Y_c} = -A_1C \cdot A_1C_2 \cdot \frac{\sin(60° - \gamma)}{\sin(\alpha - \beta)},$$

which is negative when $\alpha - \beta > 0$ and positive otherwise. The power of A_1 to ω_b is

$$\overline{A_1B} \cdot \overline{A_1Y_b} = -A_1B \cdot A_1B_2 \cdot \frac{\sin(60° - \beta)}{\sin(\alpha - \gamma)}.$$

The ratio of the powers is

$$r_{A_1} = \frac{\overline{A_1B} \cdot \overline{A_1Y_b}}{\overline{A_1C} \cdot \overline{A_1Y_c}} = \frac{A_1B_2 \cdot \sin(60° - \beta) \cdot \sin(\alpha - \beta)}{A_1C_2 \cdot \sin(60° - \gamma) \cdot \sin(\alpha - \gamma)}$$

$$= \frac{A_1P \cdot \frac{\sin 3\gamma}{\sin \gamma}}{A_1P \cdot \frac{\sin 3\beta}{\sin \beta}} \cdot \frac{\sin(60° - \beta)\sin(\alpha - \beta)}{\sin(60° - \gamma)\sin(\alpha - \gamma)}$$

$$= \frac{\sin 3\gamma \sin(60° - \beta)\sin(\alpha - \beta)\sin \beta}{\sin 3\beta \sin(60° - \gamma)\sin(\alpha - \gamma)\sin \gamma}.$$

Simplify the above fraction by the triple angle identity

$$\sin 3x = 4\sin x \sin(60° - x)\sin(120° - x)$$

to reach

$$r_{A_1} = \frac{\sin(120° - \gamma)\sin(\alpha - \beta)}{\sin(120° - \beta)\sin(\alpha - \gamma)} = r_A.$$

Now let ω_b and ω_c intersect at two points X and Y, as shown in Fig. 10.11. By the circle power theorem, A, A_1, X, Y are concyclic. The common point of intersection P defined in (2) lies on XY, the radical axis of ω_b and ω_c. Finally,

$$\overline{XP} \cdot \overline{PY} = \overline{B_1P} \cdot \overline{PB_2} = \overline{A_1P} \cdot \overline{PA_2} \Rightarrow A_1, A_2, X, Y \text{ are concyclic,}$$

and thus A, A_1, A_2, X, Y are all on a circle. The conclusion follows.

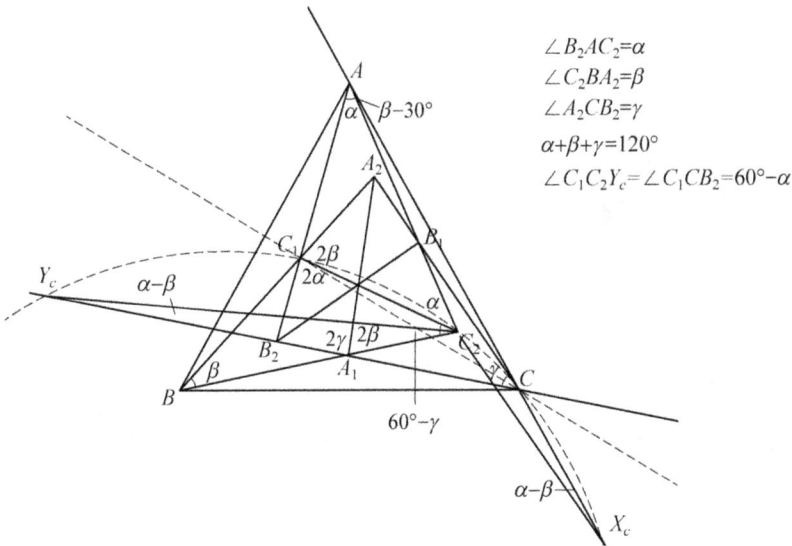

$$\angle B_2AC_2 = \alpha$$
$$\angle C_2BA_2 = \beta$$
$$\angle A_2CB_2 = \gamma$$
$$\alpha + \beta + \gamma = 120°$$
$$\angle C_1C_2Y_c = \angle C_1CB_2 = 60° - \alpha$$

Fig. 10.11